A SOURCEBOOK FOR
PHYSICAL CHEMISTRY TEACHERS

ESSAYS IN PHYSICAL CHEMISTRY

W. T. Lippincott, Editor

Members of Editorial Committee
Alan L. McClelland
Peter R. Rony
Chadwick A. Tolman

Society Committee on Education
American Chemical Society
1988

ACKNOWLEDGMENTS

The Editor and Editorial Committee acknowledge with appreciation the assistance of the following individuals and organizations:

>Sherril D. Christian
>Department of Chemistry
>The University of Oklahoma

>Gilbert W. Castellan
>Department of Chemistry
>University of Maryland

>Moses Passer
>Kenneth M. Chapman
>Sylvia A. Ware
>American Chemical Society

>ACS Society Committee on Education
>ACS Division of Physical Chemistry

Contributions from the ACS Society Committee on Education, the Committee on Corporation Associates, and the Divisions of Chemical Education, Industrial and Engineering Chemistry, and Physical Chemistry are gratefully acknowledged.

Special appreciation is due the participants, sponsors, and supporting organizations of the ACS 1984 Invitational Workshop, "Content of Undergraduate Physical Chemistry Courses," including:

>ACS Committee on Professional Training
>ACS Division of Chemical Education
>ACS Division of Industrial and Engineering Chemistry
>ACS Division of Physical Chemistry
>ACS Committee on Corporation Associates
>E. I. du Pont de Nemours and Company

Coordinator
Janet M. Boese

Graphics & Production
Leroy L. Corcoran, Production Manager
Alan Kahan, Art Director
Amy J. Hayes, Artist

Editorial
Elizabeth Wood, Supervisor
Patricia N. Rogers
Gail M. Mortenson

Essays in Physical Chemistry (soft cover, ISBN 0841214786; hard cover, ISBN 084121470-0) published by the American Chemical Society.
Copyright © 1988 American Chemical Society. Copies available from the Distribution Office, American Chemical Society, 1155 16th Street, N.W., Washington, D.C. 20036.

TABLE OF CONTENTS

Preface	Alan L. McClelland E. I. du Pont de Nemours & Co.	v
Chapter 1:	Introduction and Overview W. T. Lippincott Department of Chemistry, University of Arizona	1
Chapter 2:	Teaching Physical Chemistry G. A. Crosby Department of Chemistry and Chemical Physics Program Washington State University	5
Chapter 3:	The Coupling of Physical and Chemical Effects— The Most Interesting Things in Chemistry Chadwick A. Tolman and Nancy B. Jackson E. I. du Pont de Nemours & Co.	11
Chapter 4:	Continuity of Species in Physical and Chemical Processes Peter R. Rony Department of Chemical Engineering, Virginia Polytechnic Institute and State University	29
Chapter 5:	Building Enthusiasm for Quantum Mechanics and Statistical Mechanics Henry A. McGee, Jr. Department of Chemical Engineering, Virginia Polytechnic Institute and State University	65
Chapter 6:	Molecular Thermodynamics of Fluid Mixtures J. M. Prausnitz Department of Chemical Engineering, University of California— Berkeley	85
Chapter 7:	Polymer Examples of Thermodynamics, Statistical Mechanics, and Chemical Kinetics Leo Mandelkern Department of Chemistry, Florida State University	101
Chapter 8:	Applications of Thermodynamics and Kinetics in Inorganic Chemistry Edward L. King Department of Chemistry, University of Colorado	119
Chapter 9:	Physical Chemical Analysis of Biopolymer Self-Assembly Interactions M. Thomas Record, Jr., and Brough Richey Departments of Chemistry and Biochemistry, University of Wisconsin—Madison	145
Chapter 10:	Computer Methods in the Calculation of Phase and Chemical Equilibria Stanley M. Walas Department of Chemical & Petroleum Engineering, University of Kansas	161

PREFACE

The content and effectiveness of undergraduate physical chemistry courses have been the subject of a number of workshops and seminars in recent years. The 1980 workshop on "Cross-Fertilization between Chemistry and Chemical Engineering," sponsored by the ACS Education Commission, expressed concern about introductory physical chemistry courses and recommended that a study of the situation be made. In March 1984, the ACS Society Committee on Education (SOCED) held an invitational workshop entitled "Content of Undergraduate Physical Chemistry Courses." In attendance were chemical engineers, industrial chemists, authors of physical chemistry textbooks, and physical chemists from academe. The Committee on Professional Training and the Divisions of Chemical Education, Industrial and Engineering Chemistry, and Physical Chemistry were cosponsors. Among the recommendations of this workshop was a call to publish and disseminate a *Sourcebook for Physical Chemistry Teachers*.

The *Sourcebook* was planned "to assist and reinforce teachers in implementing the changes recommended by this workshop and by other groups such as the Committee on Professional Training and the cosponsoring divisions." It contains a wealth of examples of how former students of physical chemistry have used what they learned in their study of the subject. Industrial chemists, chemical engineers, physical chemists, inorganic chemists, polymer chemists, and biochemists all share their experiences on these pages.

If there is one underlying theme or hidden agenda here, it is a plea to help students become more confident in dealing with the physical chemistry of substances, systems, and processes. For most of the authors of the *Sourcebook*, this includes teaching methods and skills for understanding dynamic, open systems consisting of mixtures of molecules and ions that interact and react with one another. The wherewithal for doing this is available. The mathematics and reasoning skills required are accessible to the minds and abilities of undergraduates. Some very hard work and extraordinary imagination will be required of teachers. Nearly all who have examined the situation agree that improvement is needed. We hope the *Sourcebook* will catalyze this process.

Alan L. McClelland
Chairman, SOCED 1983–86

Chapter 1
INTRODUCTION AND OVERVIEW

W. T. LIPPINCOTT
Department of Chemistry
University of Arizona
Tucson, Ariz. 85721

The two ACS Workshops whose recommendations and concerns led to this volume emphasized both the importance and the perceived shortcomings of the undergraduate physical chemistry course. The report of the 1984 Workshop stated:
   The undergraduate physical chemistry course is recognized as seminal by chemists and chemical engineers. Yet, segments of both groups have voiced concern repeatedly and for some years about possible shortcomings in the course as it is customarily taught. Sizable groups of academic chemists, industrial chemists, and chemical engineers have long felt that students are not getting what they should from it.
Included in the recommendations of the 1980 Workshop is the following:
   Detailed consideration of the current content of physical chemistry courses led to agreement that physical chemistry seems to have departed from the level of usefulness it once had, particularly in providing common ground for chemists and chemical engineers and presenting the base from which chemical engineering concepts are developed. Reemphasis in physical chemistry of classical topics such as thermodynamics, kinetics, and phase equilibria would effect restitution of physical chemistry as the junction for cross-fertilization of chemistry and chemical engineering curricula. The incorporation of principles of chemical dynamics correlating chemical engineering approaches—basic concepts such as heat and mass transfer and conservation of mass—with chemical reactions and separations establishes physical chemistry as the bridge between theory and application; this merger also provides both chemists and chemical engineers with the preparation necessary for successful industrial careers.
Of particular importance are the recommendations of the 1984 Workshop.
   To emphasize and to begin implementing the key conclusions of this Workshop, participants recommend:
   1. That undergraduate physical chemistry courses and textbooks undergo the following small but significant changes:
   - That chemical dynamics be expanded to increase student familiarity with flow phenomena and with interrelationships among concentrations, space, time, and temperature.
   - That thermodynamics be applied more extensively to systems in which components are allowed to enter and/or leave the system; that additional emphasis be placed on nonideal systems, including estimation of properties, compressibility factors, activity coefficients, etc.
   - That the presentation of quantum mechanics be modified as necessary to provide a sound but compact and perhaps less mathematically distracting understanding of the material, and that a brief introduction to statistical thermodynamics be included.
   2. That, in order to provide flexibility in adding new material to the physical chemistry course, portions or all of the following topics be moved to other areas in the chemistry curriculum:
   - Electrochemistry, including electrolysis and electrode processes
   - Analytical applications of spectroscopy
   - X-ray, electron, and neutron diffraction

- Phase diagrams in complex systems
- Applications of group theory
3. That, at those institutions offering three semesters of undergraduate physical chemistry lecture, all chemistry and chemical engineering students take all three semesters.
4. That the efficient and imaginative use of computers becomes more pervasive throughout the course, including laboratory; that this be done without diminishing appreciation of the analytical dimension of the subject.
5. That plans proceed for the publication and dissemination of the *Sourcebook for Physical Chemistry Teachers*.

As its subtitle indicates, this volume is intended to be the *Sourcebook* requested in the last recommendation above. It also addresses directly and indirectly several of the other recommendations.

The focus and strategy

Each of the following chapters was written for the *Sourcebook* by chemists or chemical engineers who have extensive experience in teaching and/or applying the principles normally introduced in the undergraduate physical chemistry course.

In preparing their chapters, authors were asked to provide as many examples as possible of uses of principles, everyday applications, and motivating approaches to presentation of material. Chemical engineers were asked to show us how we can make the physical chemistry we teach more meaningful and useful to their students. Industrial chemists were encouraged to share their experiences in using basic physical chemistry principles in solving problems and in research, development, and production. Academic chemists in nonphysical chemistry areas were urged to tell us how they and their colleagues are using what they learned in physical chemistry. The hope is that the *Sourcebook* will provide a valuable collection of examples, experiences, and approaches that can enrich the perspective and increase the fund of useful and exciting knowledge of physical chemistry teachers.

Although the physical chemistry community, and especially its teachers, can take justifiable pride in the many and pervasive advances in the understanding of chemical and physical systems that have resulted from the application of physical chemistry principles during the past 20 years, it is clearly impossible to include in the introductory course more than a small fraction of what is known. In deciding how best to accommodate the diversity and complexity of student needs, subject matter expansion, and time limitations, some very basic guiding precepts are needed. In the many discussions that accompanied and followed the Workshops and follow-up seminars, three ideas surfaced repeatedly. The first is that the introductory physical chemistry course must not lose its quantitative emphasis, its problem-solving discipline, or its laboratory component. The second is that behavior of real systems, especially fluid mixtures undergoing chemical or physical change, should be treated in the course. The third is that descriptive elements must become a more significant portion of the course.

These descriptive elements might consist of at least two components. One might be described as a concept–appreciation element, something that puts both the math and the phenomena being treated in a descriptively comprehensible context, allowing the learner to see what the math is being used for and why. The other can be called a "where this takes us" element, and it normally follows the mastery of math and problem-solving skills in a unit. In this element, students discover, in descriptive terms, how the math and problem-solving skills they learned can be used to tackle problems beyond those assigned in an introductory unit. With this may come descriptions of how mathematical models are modified to accommodate additional variables, how numerical treatments of the modified models are achieved, or how successful such efforts have been.

Attitudes, applications, and dynamics

The chapters in this book offer many excellent examples that can be used to implement each of the three guiding precepts described above. Glenn Crosby sets the stage in his chapter, "Teaching Physical Chemistry," with a brief "state of the situation" analysis. Concluding that adding more real-life examples may be helpful but not sufficient, he identifies a deeper problem—the inability of many students to "abandon the microscopic imagery that has served [them] well for years, and adopt [the] macroscopic approach" of E, H, G, S and other defining parameters of the system. He offers some "attitudes, methods, and pedagogical approaches [that] have proved successful in [his] own classroom." These include several examples of how to teach concept appreciation. Crosby also makes a convincing case for truncating the number of topics in the course, and he urges that homework problems be more representative of the way science is practiced.

In their delightful chapter subtitled "The Most Interesting Things in Chemistry," Chad Tolman and Nancy Jackson look at physical chemistry as they see it "operating all around us and within us." They raise 20 questions, each dealing with an everyday application of physical chemistry, and they answer them as a teacher might do with students. Two of their questions are: Is it possible to cool your kitchen in summer by opening the refrigerator door? and How many calories from food are required for breathing each day? Tolman's verse "What Kind of Chemistry" (can respond to a smile or make a meadowlark sing) will cheer many readers.

Peter Rony's extensive and meticulously developed chapter, "Continuity of Species in Physical and Chemical Processes," is the first of four written by chemical engineers. Rony presents a unified approach to teaching the conservation-of-species equation. This powerful tool, which has been indispensable for chemical engineers for nearly 50 years, can be helpful to chemists in enriching their understanding in areas of kinetics, reactors, separations (chromatography and countercurrent techniques), diffusion-controlled processes, and electrochemistry. A thoughtfully conceived, brief presentation of this topic in the undergraduate physical chemistry course can add a new dimension of thinking and understanding in the realm of dynamic processes and open systems. It can give students a perspective that takes them beyond closed, ideal systems and provide them with skills that can enable them to deal more confidently with many of the real systems they will encounter in their work.

Quantum mechanics and real systems

No component of the undergraduate physical chemistry course has come under greater fire than the treatment of quantum mechanics. Yet no aspect of the science is more powerful or more potentially useful. Among the recommendations from the 1984 Workshop (cited above) is one that urges a more compact and less mathematically distracting treatment of quantum mechanics accompanied by a brief introduction to statistical mechanics. Such a request is far easier to make than to implement.

Henry McGee is a professor of chemical engineering with a long-standing interest in getting engineering students excited about quantum mechanics and statistical mechanics. His chapter summarizes the philosophy, the perspective, and some methods he has developed over the years. Blending quantum and statistical mechanics, and tying every development to real-world examples and applications, McGee shows us that our present methods are not far from the mark, but that they might be more effective with modest modifications.

It has been said that few engineers spark the excitement of chemists as consistently as does John Prausnitz. In his chapter, "Molecular Thermodynamics of Fluid Mixtures," he demonstrates why this is so. Using theoretical tools familiar to chemists—statistical thermodynamics, structural theory, molecular physics, and physically grounded models of systems—he develops a molecular thermodynamics that enables us to understand, calculate, and predict the properties of ideal and nonideal mixtures, including some containing molecules that differ greatly in size, and others in which chemical reactions occur. The approach developed in this chapter is a good example of something that might be presented as a "where this takes us" descriptive element in the introductory course. It also could be the centerpiece for an advanced undergraduate course.

Leo Mandelkern's prominence as a polymer chemist and author is well recognized, as are his ardent efforts to make polymer chemistry an integral part of main-line chemistry courses. His chapter, "Polymer Examples of Thermodynamics, Statistical Mechanics, and Chemical Kinetics," is written by a physical chemist for other physical chemists. It contains many definitive examples of how principles of polymer chemistry can be introduced "in a very natural way" into the introductory physical chemistry course. The thermodynamics of rubber and other elastomers, phase equilibria, and solution properties in polymer systems under various conditions are topics addressed in the context of where and how they might best fit into the teaching program.

Some ways in which inorganic chemists are using or have used principles and methods of physical chemistry are the subject of Ed King's chapter. His emphasis is on thermodynamics and kinetics, and he provides a number of interesting and novel examples involving real systems. Treatment of the thermodynamics of ammonia synthesis highlights the dependence of entropy change on the partial pressures of reactants and products and the approach to equilibrium at different pressures. Using the thermal decomposition of water as an example, King shows how "thermodynamically infeasible reactions can be made to occur as a net result of an appropriate sequence of thermodynamically feasible reactions." An especially valuable portion of his chapter deals with the correlation of equilibrium constants for successive steps of metal-ion ligation. Inclusion of the concept of cooperativity here is most interesting. Examples of the very creative work of inorganic chemists in kinetics and mechanisms of reactions complete the chapter.

How some biochemists are using physical chemistry is illustrated in the chapter by Tom Record and Brough Richey. Here, the emphasis is on understanding the thermodynamics and kinetics of noncovalent interactions between biopolymers in aqueous solutions. A key to this is recognizing that such interactions often are exchange reactions involving solutes and solvents, and that principles obtained from model gas-phase systems do not apply. Some beautiful applications involving polar–nonpolar molecule interaction and polyelectrolyte–ion–polar molecule interaction are presented and made rational in light of principles taught in introductory physical chemistry. Examples of appropriate homework and exam problems are included.

In his chapter, "Computer Methods in the Calculation of Phase and Chemical Equilibria," Stanley Walas discusses use of the computer for calculations involving phase and chemical equilibria of nonideal substances and mixtures. He features the Newton–Raphson method throughout the chapter, and he presents methods for obtaining equations of state, enthalpy and entropy, vapor-phase equilibria, activity coefficients, liquid–liquid equilibria, melt equilibria, and a relaxation method for chemical equilibria. Specific examples illustrating each method are included. The computer language used in examples is Hewlett-Packard BASIC. Material in this chapter should be immediately and widely useful in physical chemistry classrooms and teaching laboratories.

The examples and experiences compiled in these chapters come from our colleagues in industry and academe; from chemical engineers, inorganic chemists, biochemists, polymer chemists, and physical chemists. They are offered in a spirit of sharing and in the hope that much that is here will be used directly, but that the greatest value of this effort will be in the many new ideas and classroom innovations it stimulates.

Chapter 2
TEACHING PHYSICAL CHEMISTRY

G. A. CROSBY
Department of Chemistry and Chemical Physics Program
Washington State University
Pullman, Wash. 99164-4630

In spite of the lofty position that physical chemistry holds in the minds of both teachers and students, the experience in the classroom is often (one is tempted here to use the term 'always') a disaster for both. The lecturer is appalled by the test scores, and the students are depressed, demoralized, and defeated. What could have been the great year of revelation concerning the principles of chemistry has turned into a time of confusion and despair. Something went wrong, and no one seems to know why.

This analysis may be too stark, but it is based on my numerous conversations with and observations of both students and practicing chemists over the years. Too many of our graduates, even some very distinguished ones, regard the subject of physical chemistry as too hard, too abstract, and, worst of all, esoteric and irrelevant. Something is definitely amiss with our usual methods of teaching physical chemistry, and something must be done about them.

The central position of physical chemistry in the curriculum is, however, undisputed; evidence for this assertion abounds. Physical chemistry occurs at the junior year and is the watershed for the class. Students who pass the course are permitted to enroll in those senior offerings that are designated for majors; students who cannot make the grade in p-chem are destined to receive no degree in chemistry or one that 'isn't up to the standards expected in the professional program.' From the student standpoint, physical chemistry is the top of the mountain; once over the p-chem barrier, the path toward the degree is downhill. The worst is decidedly behind them. From both the professorial and the student point of view, physical chemistry is the capstone of the chemistry curriculum.

As the pages of this book reveal, there is genuine concern about the content of physical chemistry courses. Working chemists assert that the 'real world' has been left out, that the examples in the texts are not close enough to the everyday problems faced by practicing chemists. Great areas of application of the principles have been avoided or allocated so little space that their treatment is trivial. The answer, seemingly, is to put more relevant examples into the textbooks and the course. Some very creative ones are contained in this volume.

There is little doubt that many of the practical applications of the principles embodied in physical chemistry are not in the textbooks. How could they be when physical chemistry is comprised of the principles underlying all chemistry? Yes, there is need to revise the content of the physical chemistry course to improve its relevance, but content changes alone will not, in my opinion, resolve the immense problems that students confront in the classroom. The problems are deeper. Inclusion of new relevant content must also come to terms with the problem of length, for the modern text is already too big, too heavy, and too intimidating.

I believe that the malaise of the student facing physical chemistry is more fundamental than a reaction to the irrelevance of some favorite textbook examples. To understand the situation we must analyze the educational path of the typical student. From the elementary grades to the junior year in college, the student has been taught to understand nature in terms of atomic and molecular models. Virtually all explanations of physical and chemical phenomena were couched in those terms. This is the way modern students are taught to comprehend the physical world. The visualization of chemical phenomena in terms of molecular models reaches its zenith in the organic course, where structural and mechanistic arguments permeate every page of the text. Then the student, conditioned to view the world in atomistic terms, enters a physical chemistry class. Suddenly, the emphasis is no longer on molecules and structure; the point of view has become macroscopic, and only E, H, S, G, and the defining parameters of the system are considered. These quantities apply to *all* systems, and the molecular details are irrelevant; in fact, dwelling on the molecular explanations of E, H, etc. can lead to real problems, as anyone who has taught elementary statistical mechanics can verify. Thus, the student entering the physical chemistry class is asked to abandon the microscopic imagery that has been useful for years and adopt a macroscopic approach. For many the gap is too wide. Thermodynamics becomes an esoteric construct of formulas and equations devoid of meaning and relevance. The student's effort switches from trying to understand the subject to trying desperately to pass the course. The opportunity to add the principles of physical chemistry as part of the student's meaningful knowledge base is often lost in the first few weeks of the semester.

Recommendations

Teaching the principles of physical chemistry is a demanding exercise. No matter how deep the instructor's understanding or how extensive the student's level of preparation and degree of intelligence, teaching physical chemistry is fraught with difficulties. Over the years certain attitudes, methods, and pedagogical approaches have proved successful in my own classroom, and I offer them here for consideration.

Emphasize the physical systems

Physical chemistry is chemistry, and chemistry is an experimental science. Mathematics is certainly indispensable, but physical ideas about the systems must take precedence. Therefore, it is a good rule to devote as much time to definition(s) of the systems in operational terms as to the mathematical developments. Without a clear understanding of the physical problems and of the thought processes that go into defining them in algebraic terms, students are often puzzled. Frequently they confuse what is purely algebraic manipulation with physical principles, and the central concept is lost in the analysis. Students will never 'understand' physical chemistry without a crystal-clear conception of the system, the assumptions underlying the formalism, and the difference between what is physical science and what is mathematics.

Justify the content of the course

Even when the physical system has been well defined, it is difficult for students to master a topic if they are unsure of what the goal of the derivation really is. Often a topic, such as that of Gibbs' Free Energy, is introduced in a purely formal way, and the goal (the grand utility and beauty of the central concept) is revealed only at the very end. I deem this method of introducing a new idea as time-saving instruction but genuinely flawed pedagogy. How much enthusiasm can students generate to follow long concatenations of logic unless they have some sense of the purpose of the exercise? Developments of concepts and equations must have a reason. The failure of the First Law to predict anything whatsoever about spontaneity and the approach to equilibrium is worthy of much discussion before the introduction of the Second Law. Only after students have comprehended that the tendency toward equilibrium is a new topic worth their supreme effort can an instructor hope to be able to motivate them to comprehend this central idea of physical chemistry and follow the mathematical intricacies. Understanding concepts as general and universal as that of Gibbs Free Energy will never be achieved by students who have no clear conviction that such comprehension is a necessary component of an education in chemistry.

Focus on the physical meaning of numerical answers

Being able to solve problems is absolutely necessary if a student is going to understand and use the principles of physical chemistry. A perusal of a typical examination provides evidence that a certain degree of problem-solving capacity is definitely necessary to pass a standard course. Yet the ability to obtain a numerical answer with great speed and accuracy may mask a fundamental ignorance of the basic principles involved. This malady, exacerbated by examinations, is pandemic. A recommended procedure is to ask for the meaning of each numerical answer obtained. Does it make sense? Is the sign right for the process under consideration? A negative enthalpy of vaporization should certainly alert the student that an error is embedded somewhere in the calculation. If it does not, then a fundamental idea has been lost somewhere. Asking a question or two about the physical meaning of each numerical answer will reveal much about the degree of physical understanding possessed by students. It is a practice worth any instructor's effort.

Examples:

1. To make the student reflect upon the meaning of a calculation, a trivial example is to pose a numerical problem where ΔG^0 is -10 kJ for a reaction and then inquire whether, at equilibrium, the vessel contains mostly reactants, mostly products, or significant concentrations of both reactants and products.

2. This standard question offers many other intriguing possibilities. For the listed reactions, indicate whether the change in the thermodynamic quantity is $<0, 0, >0$. Assume standard states (unless specified), 298 K, and ideal gases.

	ΔG	ΔH	ΔS
$CaCO_3(s) \rightarrow CaO(s) + CO_2(g)$	$<0, 0, >0$	$<0, 0, >0$	$<0, 0, >0$
$N_2(10 \text{ atm}) \rightarrow N_2(20 \text{ atm})$	$<0, 0, >0$	$<0, 0, >0$	$<0, 0, >0$
$CH_4 + 2O_2 \rightarrow CO_2 + 2H_2O$	$<0, 0, >0$	$<0, 0, >0$	$<0, 0, >0$

3. A simple question that causes a class of engineers to scratch heads for a long time is based on carry-over from one class to another—something students often find difficult: The substance (CH_3—CH_2—O—CH_2—CH_3) was thrown on a warm surface. A process occurred. Assign ($<0, 0, >0$) for ΔG, ΔH, ΔS for the process. (*Hint:* What happened?)

Build mathematical bridges

Much of physical chemistry is concerned with the process of casting physical problems into mathematical language. For most students this is not easy; for some of them it is a major accomplishment. This inherent difficulty is compounded by the fact that the mathematical formalism has often been assimilated in another course (calculus) and in a foreign notation. What is to the professor a simple Taylor Series is often not recognized by the student because of the different notation: P, V, T do not look like x, y, z, and therefore the correspondence can be easily missed by a student who was uncertain of the mathematical ideas presented in the calculus class in the first place. In my opinion, time spent by the chemistry professor in delineating the correspondence is time well spent. Why allow a trivial notational switch to block understanding of a topic as important and interesting as real gases? Professorial time spent perusing the current mathematics texts would also be rewarding. Requiring the students to build all the mathematical and notational bridges is not only inefficient, but it is also unfair. Mathematical notation has progressed and changed with the years. What students are now exposed to is not as closely related to our thought processes as the older mathematical developments were. *We* are out of touch, not the students.

Reinstate demonstrations

Of all these ways of improving the teaching of physical chemistry, the use of demonstrations is perhaps the most likely to be ignored. Yet there is ample evidence that all students do not learn in the same way. Some remember a demonstration (a light glowing, a pen wiggling, an explosion) long after the details of a discipline have passed into the subconscious. It can be said that the lab is where the student sees what is going on, where the physical knowledge of systems is acquired. There are lots of good physical chemical principles that students do not experience in the laboratory, however, and many of these do lend themselves to simple dramatic demonstrations (*vide infra*).

For an introduction to E, the internal energy, one can make marvelous use of the proverbial H_2/O_2 explosion in a balloon. What was the temperature *before* the explosion? *After* the explosion? What was the function of the match? Was $q >0$, 0, <0? Was w nonvanishing? Where was the released energy stored? Then, one can pass to thought experiments. Suppose we had exploded the mixture in a steel, insulated bomb. Would q change, etc.? One good demonstration can underscore what the intent should be—to focus attention on the physical system before becoming involved in mathematics.

Few of us do any lecture demonstrations at all in upper level courses. The prevailing opinion seems to be that the student will achieve a good physical knowledge in the labs (which is very doubtful) or that the descriptions in the text are adequate substitutes for direct observation. I disagree with the conventional wisdom.

With the growing accessibility of computers in the classroom there is an obvious substitute for the real demonstrations—computer simulations. Computers can furnish visual images that are superior to those of the chalkboard even if they are used only to draw density maps.

Finally, if demonstrations are not feasible and computer simulation is too difficult to employ, one can always resort to the thought experiment—a graphic description of what a real system actually does when subjected to the defined constraints. Going from a real system to a mathematical model of an ideal system is the logical path that every student must take to comprehend physical chemistry. Visual supplements can help students establish those links.

Design balanced examinations

As every experienced teacher knows, the style and emphasis of examinations usually define what students learn in the course. If no call is made for the derivations of pertinent formulas on the examinations, then the students will not attempt to be able to reproduce them. If one does not ask for a critical analysis of the assumptions underlying any equations, few students will bother to study them critically. If the tests consist entirely of numerical calculations, then that is what the students will learn to do. If we want students to focus attention on the meaning of the quantities calculated, then we must construct examination questions that probe for the physical meaning of the numerical answers.

Constructing questions demanding numerical answers is easy; the questions are also easy to grade. Formulating questions that ask for discrimination and judgment is difficult. Yet the latter type of question can reveal a depth of understanding that is not really possible with numerical problems. A few examples of nonnumerical problems are:

1. Provide a list of thermodynamic data in tabular form and ask the students to correlate the entries with a list of substances.
2. Ask the students to estimate the magnitudes of measurable quantities.
3. Plot equilibrium constants as a function of temperature and ask questions that leave the students no alternative but to think of the physical meaning of the quantity under discussion, which is the intended result. An example of a series of nonnumerical questions is shown in Figure 1.

From a consideration of Figure 1, answer the questions below:
(a) Which process(es) could not represent vaporization?
(b) Which process has the smallest absolute value for the standard enthalpy change?
(c) Which process(es) has (have) a temperature-independent enthalpy change?
(d) For process (C), ΔH^0 at T_1 is (lower than, the same as, higher than) ΔH^0 at T_2. (Circle one.)

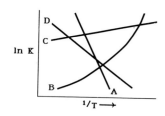

Figure 1. Plot of ln K vs. 1/T for a series of processes.

These are only a few examples of the way one can probe the students' knowledge without demanding a calculation. I have always included a liberal mix of such questions on examinations. They are difficult to write, they are not easy to grade, but they are phenomenally effective in directing the students' attention to the physical essence of what is often only mechanically calculated.

Design homework problems more representative of the way science is practiced

Many homework problems are contrived. The answer is known and the student is asked to calculate the observations. For instance, once the geometry of a molecule is known, the expected microwave spectrum is easy to compute. Yet the process by which the geometry is determined is the precise opposite. One starts with the microwave spectrum and 'works backward,' i.e., one goes from the spectrum to the model and not from the model to the spectrum. Going from the observed lines to the model is a much more difficult problem, not only because the arithmetic is tedious, but also because the answer depends upon the resolution of the lines, the type of statistical analysis employed, the number of lines observed, etc. Nonetheless, this is exactly the kind of problem that a working scientist faces. In the past, constructing problems that more closely resembled the real world was time-consuming, both for the professor and for the student. A single problem could take hours to solve, even when the method was clearly understood. Today, however, the situation is not so grim. One can program a computer to simulate data and, by a creative use of graphics and random number generators, produce data with striking verisimilitude. The student also has access to statistical and computational power to aid in the solution of such problems. Homework problems of the type described here could and should be assigned as projects to groups of student. One seldom works alone in the real world of chemistry.

Truncate the number of topics in the course

In response to the need for new and different material in the physical chemistry course, it appears contradictory to issue a call for less instead of more. In my opinion, however, to focus exclusively on coverage and content without considering the size of the books already in use would only add to the problems we now face. In short, not only must we change the content, but we must achieve more depth. It is apparent that the only variable left is breadth of coverage, and something must be eliminated.

To focus on coverage in a course, any course, is to miss the point of education altogether. Our aim should be to produce students who know some of the basic principles of all chemistry, who have confidence in what they know, and who understand instinctively how to expand their knowledge. Moreover, we should try to present the subject so that the student leaves the course knowing that the subject is basic, that it can be learned, that it is practical, and that there is more to be learned and much more to be discovered. In short, we should aim to produce chemistry graduates who are problem solvers and problem generators, not just students who can solve problems. Viewed from this angle, what is covered in the physical chemistry course is far less important than how the subject is taught and what behavior the students exhibit after leaving the course.

Summary

In this short essay I have attempted to focus on the pedagogical problems that arise in the teaching of physical chemistry, some possible ways of resolving them, and what I believe the goals of the course ought to be. Physical chemistry is an important component in the chemistry curriculum. Its potential for deepening the students' knowledge of and appreciation for chemistry is immeasurable. Too many of our students never really achieve the depth of understanding commensurate with their abilities, and far too few derive any enjoyment whatsoever from the course.

Chapter 3

THE COUPLING OF PHYSICAL AND CHEMICAL EFFECTS

The Most Interesting Things in Chemistry

**CHADWICK A. TOLMAN AND
NANCY B. JACKSON**
*Central Research and Development Department
E. I. du Pont de Nemours & Co.
Wilmington, Del. 19898*

Physical chemistry in its broadest sense involves the interaction of physical and chemical effects—e.g., the interactions of light, heat, kinetic energy, and electrical, magnetic, and gravitational fields with chemical reactions, structures, and properties—and is thus operating all around us, and within us. Helping students to see this, to think in this way, and to be struck with a sense of awe and wonder at the world around them should be, in our view, a major goal of an introductory course in physical chemistry.

This chapter has been written to help physical chemistry teachers in that direction. We suggest that for these exercises the class be divided into groups of two to four students each. Each group is given a question and allowed 10 to 15 minutes in class to discuss it and arrive at their best collective answer, which is then presented to the class by one member from each group. Students are then given additional information and references for further reading; they are asked to report back to the class on what they find, pointing out what new things they have learned as a result of their further reading. The following is a list of possible topics. It might be interesting to have a few groups working on questions from each topic in order to compare their answers.

- Refrigerators and air conditioners
- Jogging and heat
- The work of breathing
- Air temperature and altitude
- Hurricanes
- Boiling eggs at 10,000 ft
- Candles burning in zero gravity
- Microwave ovens
- The stretching and oxidation of rubber bands
- Glues and adhesives
- Paints and surfactants
- Home runs and humidity
- Graphite and diamonds
- Antique window glass
- Platinum and hydrogen
- Shape selectivity in zeolites
- Substrate selectivity in zeolites
- Types of continuous reactors
- Electric eels
- Chemical oscillations and biorhythms
- Epilogue

Refrigerators and air conditioners

Question
Is it possible to cool your kitchen in summer by opening the refrigerator door? How is an air conditioner different?

Answer
No. Opening your refrigerator door will not keep your kitchen cool.

In a thermodynamic refrigeration cycle, heat is absorbed at a low temperature and is rejected to the surroundings at a higher temperature. The "surroundings" is your kitchen in the case of the refrigerator (check the bottom and back of your refrigerator for heat pouring out) and is the outside in the case of your air conditioner. This definition of the surroundings is the thermodynamic difference between an air conditioner and a refrigerator.

A refrigeration system takes advantage of the fact that in order to evaporate a liquid, heat must be added. (This phenomenon is seen every time water is boiled.) When a refrigerant such as Freon® is evaporated, it takes the energy necessary for vaporization from itself and its surroundings, lowering the temperature of both.

The refrigeration cycle consists of the evaporation of liquid by rapid expansion (i.e., rapid lowering of pressure), which gives a corresponding decrease in temperature. (If you have ever released a dry CO_2 fire extinguisher, you know how cold a rapidly expanding gas can get.) This cooling is used to keep the refrigerator cold. In the next step of the cycle the refrigeration fluid (coolant) is compressed, causing the temperature to rise enormously. This is step BC on Figure 1. The coolant, which at this

Figure 1. The refrigeration cycle.

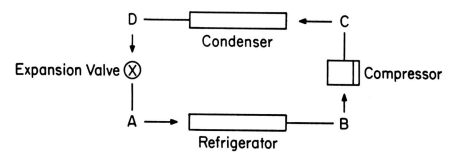

point is a gas because of the high temperature, is then allowed to cool and condense in a heat exchanger. Physically, this is accomplished by channeling the coolant gas through small-diameter tubes and blowing room air over the tubes with a fan. (This is the warm air you feel blowing out of the bottom of your refrigerator.) This way heat is rejected to the surroundings at a higher temperature than the one at which it was absorbed. After the coolant gases condense (step CD), they go through the expansion/evaporation process and the cycle repeats itself.

The *ideal* refrigeration cycle is just the opposite of a Carnot heat engine. It is considered to be reversible, and therefore the most efficient possible process. A look at the pressure-enthalpy diagram (Figure 2) shows that more heat is liberated in the condenser step (CD) than in the heat absorption step (AB).

Figure 2. The ideal refrigeration cycle.

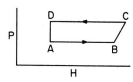

This may also be seen in the following equation for a Carnot refrigeration cycle:

$$\frac{Q_H}{Q_R} = \frac{T_H}{T_R}$$

Q_H = amount of heat given off at condenser
Q_R = amount of heat absorbed at refrigerator
T_H = temperature at which heat is rejected (temperature of coolant gas entering the condenser)
T_R = refrigerator temperature

This equation says that the absolute value of the amount of heat given off at the high temperature compared with the amount absorbed at the lower temperature is equal to the ratio of the two temperatures. This equation also supports the conclusion that the heat given off in a refrigeration cycle is greater than the heat absorbed. Therefore, opening your refrigerator door will only cause a net warming of your kitchen, not a cooling.

References
(1) Smith, J. M.; Van Ness, H. C. *Introduction to Chemical Engineering Thermodynamics*, 4th ed.; McGraw-Hill: New York, 1987; Chapters 5 and 12.
(2) Bent, H. A. *The Second Law;* Oxford University Press: New York, 1965.
(3) Haywood, R. W. *Analysis of Engineering Cycles*, 2nd ed.; Pergamon Press: Oxford, 1975; Chapters 5 and 10.

Jogging and heat

Question

A jogger consumes 100 Calories (100 food calories) per mile. Assuming that sweating is her only mechanism to maintain her body temperature at 98.6 °F, how much water (in liters) must she evaporate in a 26-mile marathon? (Remember: 1 food Calorie = 1 chemical kcal.)

Answer

4.5 L.

For water,

ΔH_{vap}, 98.6 °F = 10.4 kcal/mol

Therefore,

$$\frac{1 \text{ mol H}_2\text{O}}{10.4 \text{ kcal}} \times \frac{18 \text{ g H}_2\text{O}}{\text{mol H}_2\text{O}} \times \frac{1 \text{ L}}{1000 \text{ g H}_2\text{O}} \times \frac{100 \text{ kcal}}{\text{mi}} \times 26 \text{ mi} = 4.5 \text{ L}$$

For every mole of water evaporated off the skin of the jogger, 10.4 kcal of heat are also removed. Since 100 kcal of heat are produced for every mile, it is a simple matter of algebra and keeping track of units to estimate the amount of water required to keep the jogger cool.

The enthalpy of vaporization of water may be found in most physical chemistry textbooks, especially ΔH_{vap}, 298 K, which is slightly higher than ΔH_{vap}, 98.6 °F.

The heat of vaporization at a variety of temperatures may be obtained from the following two sources:

References
(1) *CRC Handbook of Chemistry and Physics*, 65th ed.; Weast, R. C., Ed.; CRC Press: Boca Raton, Fla., 1984.
(2) *Perry's Chemical Engineer's Handbook*, 64th ed.; Green, D. W., Ed.; McGraw-Hill: New York, 1984.

The 4.5 L (more than a gallon!) estimated here is clearly too large, since some air cooling no doubt also takes place; the calculation does, however, emphasize the need to replace the substantial amounts of water that can be lost during vigorous exercise.

The work of breathing

Question

How many calories of food energy are required for breathing each day, assuming that one inhales and exhales an average of 0.5 L of air 15 times/min against 1-atm pressure? (*Hint:* 1 L-atm = 24.2 chemical calories.) If 30% of the O_2 inhaled is consumed, what percentage of the energy generated from the O_2 consumed is used to operate the lungs? (Assume O_2 is used to oxidize sucrose to CO_2 and H_2O.)

Answer

Since the work is done against a constant 1-atm pressure, the work per breath is just the volume change times the pressure, of 0.5 L-atm for each exhalation and 0.5 L-atm for each inhalation.

$$W = 2 \times 0.5 \text{ L-atm} \times \frac{15}{\text{min}} \times \frac{60 \text{ min}}{\text{h}} \times \frac{24 \text{ h}}{\text{day}} \times \frac{24.2 \text{ cal}}{\text{L-atm}} = 523 \text{ kcal/day}$$

Since heat cannot be converted to work with 100% efficiency, something more than 523 Calories will actually be required to do the work of breathing.

Ambient air is 21% O_2 by volume, and a mole of an ideal gas at 1-atm and 37 °C (normal body temperature) is 22.4 × 310/273 = 25.4 L/mol. The number of moles of O_2 consumed per day is

$$0.3 \times 0.5 \text{ L} \times 0.21 \times \frac{15}{\text{min}} \times \frac{60 \text{ min}}{\text{h}} \times \frac{24 \text{ h}}{\text{day}} \times \frac{1 \text{ mol}}{25.4 \text{ L}} = 26.8 \text{ mol}$$

The heat of combustion of sucrose is 1350 kcal/mol (*1*)

$$C_{12}H_{22}O_{11} + 12O_2 \rightarrow 12CO_2 + 11H_2O$$

or 112.5 kcal/mol O_2 consumed. (Most organic compounds release about 100 kcal/mol O_2 consumed in converting them to CO_2 and H_2O.) Therefore 26.8 mol O_2 × 112.5 kcal/mol O_2 = 3014 kcal; 523 kcal is 17% of this amount.

References
(1) *Handbook of Chemistry and Physics*, 41st ed.; Hodgman, C. D., Ed.; Chemical Rubber Publishing: Cleveland, Ohio, 1959–60; p. 1920.
(2) Nadel, E. R. *Am. Sci.* **1985**, *73*, 334.

Air temperature and altitude

Question

Why does the temperature tend to decrease about 6°/km with increasing altitude, at least up to about 10 km above sea level (in the troposphere)? Above that, in the stratosphere (about 10–50 km) the temperature increases with altitude. (*Hint:* The prefix *tropo* means turning or changing, and *strato* means bed or layer.)

Answer

The temperature in the atmosphere is controlled by the amount and type of solar radiation absorbed at each altitude, and by the heat transferred by conduction and convection. Near the Earth's surface, the atmosphere is warmed by solar radiation absorbed by the Earth. Therefore the temperature is highest near the Earth's surface. As this leaves the warm air underneath and the cooler air on top, the troposphere experiences vertical mixing and turbulence, which is most often referred to as weather.

The stratosphere experiences no vertical mixing because the air temperature there increases with increasing altitude. The ozone in this layer absorbs ultraviolet radiation from the sun, causing the air to warm, especially at higher altitudes where the UV radiation is more intense. This creates a permanent temperature inversion—the term used to refer to warm air over cold.

Everyone knows that warm air, with its lower density, tends to rise over cold air. However, in our atmosphere, a pressure drop is associated with increasing altitude. Warm air will expand during its rise as the pressure drops. Because air is a poor conductor of heat, as it rises its temperature will drop because of adiabatic expansion. For an ideal gas this temperature change may be calculated from the hydrostatic equation

$$\frac{dP}{dh} = -g\rho$$

where

P = pressure
h = height
g = acceleration due to gravity, 9.8 m/s^2
ρ = density of air

and the adiabatic expansion equation for an ideal gas

$$\frac{T}{T_0} = \left(\frac{P}{P_0}\right)^{\frac{R}{C_p}}$$

T = temperature
R = gas constant
C_p = heat capacity of air

The *adiabatic lapse rate*, or the rate at which the temperature decreases with altitude, may be derived as

adiabatic lapse rate = $-gM/C_p$

where M = average molecular weight of air (about 28.8 g/mol).

For dry air, this rate is -9.8 K/km. Since water has a much higher heat capacity than air, it reduces the cooling effect of expansion and reduces the adiabatic lapse rate to about -6 K/km. This figure closely represents the average decrease in temperature found in the Earth's troposphere. The actual cooling rate in the atmosphere can be even less, due to the effects of rain (water condensation releases heat) and other factors.

As a result of the adiabatic lapse rate, the temperature at the bottom of the Grand Canyon can be over 100 °F when the temperature on the rim, a mile above, is a pleasant 75°.

References

(1) Moore, J. W.; Moore, E. A. *Environmental Chemistry;* Academic Press: New York, 1976, p. 183.
(2) Baum, R. M. *Chem. Eng. News* **1982,** *60*(37), 21.

Hurricanes

Question

Explain as many of the following observations about hurricanes as you can:
1. The pressure at the storm's center may be as much as 90 mb lower than that of the surrounding atmosphere (1000 mb = 1 bar).
2. The top layer of the ocean over which the storm passes cools off by about 2 °C.
3. Hurricanes are most frequent in the northern hemisphere in late summer and require a minimum ocean surface temperature of 27 °C.

4. The circular winds travel counterclockwise in the Northern Hemisphere and clockwise in the Southern, and they are not found in a band of 5–8° on either side of the equator.

Answer

A hurricane is a type of heat engine in which the heat energy of seawater is converted to kinetic energy in the form of strong winds above 120 km/h. A hurricane has an efficiency of only about 3% as a heat engine, yet its effects can be awesome. In 1281 a powerful hurricane (commonly called a *typhoon* in the Pacific) destroyed a large Chinese naval invasion force on its way to Japan, killing about 100,000 men. The Japanese called it *Kamikaze*, meaning "divine wind," because it saved their homeland.

Hurricanes require a certain minimum ocean water temperature (27 °C) and are therefore most frequent in late summer in the Northern Hemisphere. A hurricane begins in an area of reduced pressure—caused by local heating—recognizable by the presence of closed isobaric lines on a pressure map. Air moving in toward the low-pressure area is deflected to the right in the Northern Hemisphere by coriolis forces (a coriolis force is a force proposed by a person in a rotating coordinate system to explain why objects moving toward the rotational axis don't appear to travel in a straight line), resulting in a counterclockwise air flow. Moist air rising in a column in the center reaches saturation as it rises and cools. The heat transferred to the air by the condensing water adds buoyancy and upward velocity to the air column, further reducing the pressure at sea level. Hurricanes can involve a mass of air 10 km deep and 1000 km in diameter. They are not found near the equator because the coriolis force vanishes there.

Reference

(1) Lugt, H. J. *Vortex Flow in Nature and Technology;* John Wiley and Sons: New York, 1983; Chapter 11.

Boiling eggs at 10,000 ft

Question

If the barometric pressure falls 3% for each 1000-ft increase in elevation, what is the boiling point of water at 10,000 ft? If the activation energy for cooking an egg is 30 kcal/mol, how long will it take to boil an egg at 10,000 ft to the same degree of hardness as a 3-min egg at sea level? (Assume that cooking an egg is a first-order rate process.)

Answer

9 min

First, calculate the barometric pressure P_1 at 10,000 ft, assuming that the pressure P_0 at sea level is 760 torr.

$$P_1 = P_0(0.97)^{10} = 560 \text{ torr} \tag{1}$$

Second, find the boiling point of water at this pressure. This may be done in more than one way. One is to use the Clausius-Clapeyron equation.

$$\frac{dP}{dT} = \frac{\Delta H_{vap}}{T \Delta V_{vap}} \tag{2}$$

Assuming that

$$\Delta V_{vap} = \frac{RT}{P}$$

and using the formula

$$dx/x = d(\ln x)$$

Equation 2 may be rearranged to give

$$\frac{d(\ln P)}{dT} = \frac{\Delta H_{vap}}{RT^2} \tag{3}$$

Assuming that the ΔH_{vap} does not change significantly with temperature, Equation 3 may be integrated to a useful form.

$$\ln \frac{P_1}{P_0} = \frac{-\Delta H_{vap}}{R} \left(\frac{1}{T_1} - \frac{1}{T_0} \right) \tag{4}$$

Substitute the following values into Equation 4 and solve.

$P_0 = 760$ torr
$P_1 = 560$ torr
$\Delta H_{vap} = 9.72$ kcal/mol (540 cal/g)

R = 1.987 cal/mol K
Since T_0 = 373 K, T_1 = 363 K or 90 °C

A quick way of finding this temperature is to use a table listing the vapor pressure of water at various temperatures. References 1 and 2 include this information.

The third step involves the kinetics of cooking the egg. Since the process is first order, the appropriate equation is

$$-d[A]/dt = k[A] \qquad (5)$$

The variable [A] can be thought of as representing the concentration of uncooked egg in the entire egg at time t. When the egg is raw, $[A] = [A]_0$.
Integrating Equation 5 and rearranging gives

$$[A] = [A]_0 e^{-kt} \qquad (6)$$

To get the same degree of hardness means that the reaction must be carried out to the same extent. Therefore,

$$k_1 t_1 = k_0 t_0 \qquad (7)$$
or $t_1/t_0 = k_0/k_1$

Since the activation energy for the reaction is 30 kcal/mol and

$$k = A \exp(-E_a/RT) \qquad (8)$$

$$t_1/t_0 = k_0/k_1 = \exp\left[\frac{-30,000}{1.987}\left(\frac{1}{T_0} - \frac{1}{T_1}\right)\right]$$

t_1/t_0 = 3 or t_1 = 9 min

(That's why high-altitude campers fry their eggs for breakfast, rather than boiling them.)

References
(1) *CRC Handbook of Chemistry and Physics*, 65th ed.; Weast, R. C., Ed.; CRC Press: Boca Raton, Fla., 1984.
(2) *Perry's Chemical Engineer's Handbook*, 6th ed.; Green, D. W., Ed.; McGraw-Hill: New York, 1984.

Candles burning in zero gravity

Question
Can you have a candlelight dinner in an orbiting space station? How does a candle burn, and how will that be affected by weightlessness?

If a candle burns at a rate of 14 g/h, what is the rate of O_2 consumption in mol/s? If the same candle is put in a capped 1-L bottle, how long will it burn (with gravity) before it goes out? (*Hint:* Assume that the wax is pure saturated hydrocarbon and that it burns to CO_2 and H_2O.)

Answer
A candle burns as wax, melted by the heat of the flame, travels up the wick by capillary action, and is vaporized. Combustion takes place at the flame edge, where the fuel and oxygen meet. The hot gas created by the flame is less dense than the surrounding cooler air. As the hot gas rises, cool air rich in oxygen comes in to replace it, giving a convection flow pattern like that shown in Figure 3.

On Earth, oxygen is supplied to the burning fuel both by convection and from diffusion across the concentration gradient near the flame. Under weightless conditions the only mechanism to supply oxygen is diffusion, which is much too slow, and a bright flame cannot be maintained for more than a fraction of a second!

Diffusion is described by Fick's law

$$J_x = -D\frac{dC}{dx}$$

Figure 3. Flow pattern of air around a candle.

where J_x, the flux (mol/cm² s) in the x direction, is proportional to the concentration gradient dC/dx (mol/cm³·cm). The proportionality constant D is the diffusivity, which has units of cm²/s. The minus sign means that if the concentration is increasing as x increases, the flow will be in the negative x direction, i.e., toward lower concentration.

By considering the flows into and out of a volume element, it is possible to derive an equation for the rate of change of concentration. It is a partial differential equation because the concentration of a species may depend on both time and position.

$$\frac{\partial C(x,t)}{\partial t} = D\frac{\partial^2 C(x,t)}{\partial x^2} + \dot{C}$$

Here $\dot{C}$ is the rate of change of concentration as a consequence of chemical reaction. In our problem, $C = 0$ inside the flame since all reaction occurs at the flame boundary.

The concentration of oxygen outside the flame in three dimensions can be written

$$\frac{\partial C}{\partial t} = D\nabla^2 C + \dot{C}$$

where $\nabla^2 = \frac{\partial^2}{\partial x^2} + \frac{\partial^2}{\partial y^2} + \frac{\partial^2}{\partial z^2}$

or

$$\frac{\partial C}{\partial t} = D \frac{1}{r^2}\frac{\partial}{\partial r}\left[r^2 \frac{\partial C}{\partial r}\right] + \dot{C} \text{ in spherical coordinates.}$$

This equation could be solved, given an initial profile of oxygen concentration and a set of boundary conditions (e.g., $C = 0$ at a distance $r = a$ from the center of the flame and $C = C_0$ at a very large distance.) Instead, we will make some estimates based on the dimensions of the relevant physical parameters.

The rate at which O_2 can be supplied by diffusion (R_{diff}) will be proportional to the diffusivity, to the concentration of O_2 outside the flame, and to the size of the flame. By looking at the dimensions of the variables, we see that R_{diff} (mol/s) $\sim$ D (cm^2/s) $\times$ C_0 (mol/cm^3) $\times$ a (cm). The diffusivity of O_2 in N_2 at one atmosphere and room temperature is about 0.2 cm^2/s. (Diffusivity increases with temperature, but not strongly.) The concentration C_0 of O_2 in air is about 0.01 mol/L ($\sim$25 L/mol air which is $\sim$20% O_2) or 10^{-5} mol/cm^3; the radius of a candle flame is about 0.5 cm.

Then $R_{diff} \sim 0.2$ (cm^2/s) $\times 10^{-5}$ (mol/cm^3) $\times 0.5$ cm
$\sim 0.1 \times 10^{-5}$ mol/s

The chemical reaction for combustion is as follows:

$(CH_2)_n + 1.5nO_2 \rightarrow nCO_2 + nH_2O$

Fourteen grams of saturated hydrocarbon is 1.0 mol of CH_2 groups, and requires 1.5 mol of O_2 for complete combustion. (The hydrogens H—$(CH_2)_n$—H at the ends of the chain are neglected.) The rate of O_2 consumption will be

R = 1.5 (mol/h) $\times$ 1 h/3600 s = 0.4×10^{-3} mol/s

Thus we see that the rate of normal burning is two orders of magnitude larger than the rate at which O_2 can be supplied by diffusion!
Since a liter of air contains

1 L air $\times$ 1 mol/25 L air $\times$ 0.2 mol O_2/mol air = 0.008 mol O_2

this amount of O_2 with normal gravity will last about

$t = 0.008$ mol/0.4×10^{-3} = 20 s

Experiment: Put a burning candle in a 1-L jar (the short votive candles are best); cap it, and measure the time before the candle goes out. Repeat the experiment (preferably in a darkened room), but drop the bottle and watch the flame. You can either catch the bottle or drop it onto a soft pillow so that it doesn't break.

References
(1) Bird, R. B.; Stewart, W. E.; Lightfoot, E. N. *Transport Phenomena*; J. Wiley & Sons: New York, 1960.
(2) Tinoco, I., Jr.; Sauer, K.; Wang, J. C. *Physical Chemistry*, 2nd ed.; Prentice Hall: Englewood Cliffs, N.J., 1985.
(3) Penner, S. S. *Chemistry Problems in Jet Propulsion*; Pergamon Press: New York, 1957; p. 276.
(4) Kumagai, S.; Sakai, T.; Okajima, S. *13th International Symposium on Combustion*; The Combustion Institute: Pittsburgh, 1971; pp. 779–85.

Microwave ovens

Question
How does the operation of a microwave oven depend on the H—O—H angle of the water molecule? Why does cooked meat heat more evenly than raw meat in a microwave oven? What causes the electrical arcing from metal left in a microwave?

Answer
Microwaves (λ = 1 mm–30 cm) lie between the low-energy radio frequencies and the higher radiation frequencies of heat and light. This relatively low-energy radiation corresponds to the rotational energy of many molecules. The molecules absorbing the energy must have a permanent dipole moment, which on rotation appears to an observer in the plane of rotation as a fluctuating dipole. Since H_2O is a bent molecule, it possesses a permanent dipole moment and can absorb microwave energy. Carbon dioxide is also a triatomic molecule; however, it is linear, and therefore has no permanent dipole moment. The absorption of microwave energy by a water molecule causes it to rotate. This energy is translated into heat by the friction of the energized movements of the molecules against one another, and this heat cooks the food.

When raw meat is cooked, the proteins denature, which means that the proteins unfold out of their tertiary structure. Protein chains exist in nature in unique folded configurations. Interactions (H-bonding, disulfide bonds, etc.) between functional groups far apart from one another in the amino acid sequence hold the protein chain in a specific three-dimensional configuration, which minimizes its overall energy. Heat causes the proteins to vibrate and, at a certain temperature (usually under 100 °C), gives enough energy to destroy the protein tertiary structure. Denaturing releases water and fat molecules that are often tucked into the folds of the protein chain. This irregular releasing of packets of water causes local heating and, therefore, some parts of the meat heat faster than others. Since the proteins of cooked meat are already denatured, cooked meat may be warmed more evenly than raw meat.

Microwave bombardment of a metal provides enough energy to knock out electrons, causing a surface charge to develop. This charge causes a current to flow on the surface and, in seeking a ground, causes arcing within the microwave oven. Metal charging with microwave energy is strictly a surface phenomenon, as no fields, electric or magnetic, may be found inside the metal.

References
(1) Stuchly, M. A.; Stuchly, S. S. *Proc. IEEE-A* **1983**, *130*, 467.
(2) Atkins, P. W. *Physical Chemistry*, 2nd ed.; W. H. Freeman: New York, 1982; pp. 566–67, 576–79.

The stretching and oxidation of rubber bands

Question
Does a stretched rubber band contract or expand when heated? Explain. Why does a stretched rubber band deteriorate much faster than a relaxed one?

Answer
A stretched rubber band will *contract* when heated. Oxidation, which causes deterioration, is faster when the band is stretched. In order to understand both these phenomena, it is essential to know something about the chemical composition and structure of rubber. Rubber bands are made primarily from butadiene-styrene rubber (BSR). But all rubbers, including BSR, share several structural elements that give them their unique properties.

First, all rubber materials are composed of long-chain polymers. Natural rubber, cis-1,4-polyisoprene, is one example.

$$\left[\begin{array}{c} -CH_2 \\ C=CH \\ CH_3 \end{array} \begin{array}{c} HC_2-CH_2 \\ C=CH \\ CH_3 \end{array} \begin{array}{c} CH_2- \\ \\ \end{array}\right]_n$$

cis-1,4-polyisoprene

The long chain is an essential structural characteristic; the coiling and uncoiling of these chains give the rubber its elasticity.

Second, the long chains are attached to one another by bonds between chains or cross-links. Natural rubber is vulcanized (heated with sulfur) to form cross-links with sulfur bridges (Figure 4). The cross-linking bonds cause the polymer to recover its initial shape after being stretched.

Third, rubber polymers characteristically have a large amount of remaining unsaturation, i.e., double bonds, along their chains.

In a rubber, the barriers to rotation about the carbon-carbon single bonds in the chains are low, and many configurations are possible. If a rubber band is stretched, the chains extend easily to very large deformations. At the extended state there are fewer possible configurations that the bonds can take; hence the entropy is less. Heating a rubber band causes an increase in the number of possible conformations by making it easier for higher energy conformational states to be populated. This increase in randomness (with a corresponding increase in entropy) is what produces a retractive force, pulling the polymer into a tighter ball. A typical rubber band will contract up to an inch when heated to 150 °C.

Deterioration of a rubber band is caused by the breakage of bonds in the polymer backbone or the additional formation of bonds between chains that occurs when oxygen and/or ozone react with the double bonds. Although the double bonds are somewhat sensitive to oxidation by O_2, they are many times more reactive toward

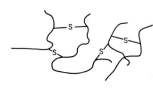

Figure 4. Formation of cross-links in heated natural rubber.

ozone. Therefore, despite the fact that ozone is found in the atmosphere at a concentration level of less than 0.1 ppm, it is the primary oxidant responsible for the deterioration of rubber.

Ozone first interacts with a rubber band by adsorbing on the surface, forming a thin oxidized layer that protects it from further ozone attack. However, when the rubber band is stretched, microscopic cracks and breaks expose fresh surfaces that are susceptible to further attack. These new points of exposure cause more chain scission and/or cross-linking and lead to the deterioration of the rubber band. Therefore, a stretched rubber band will deteriorate faster than an unstretched one. Bond scission reduces the average length of the chains, which may not change the polymer much from a chemical standpoint, but can have a profound effect on its mechanical properties. Excessive cross-linking of a rubber band in a stretched position may cause it to "set," meaning that it will not regain its shape when released.

Antioxidants such as saturated quinolines may be added to rubber to prevent ozone attack and degradation. A good antioxidant reacts with both the ozone and the rubber, shielding the new cracks created by stretching from further ozone attack.

References
(1) Jellinek, H.H.G. In *Fracture Processes in Polymeric Solids;* Rosen, B., Ed.; John Wiley: New York, 1964.
(2) Gent, A. N. *J. Appl. Polymer Sci.* **1962,** *6,* 442.
(3) Braden, M.; Gent, A. N. *J. Appl. Polymer Sci.* **1962,** *6,* 449.
(4) Bailey, P. S. *Chem. Revs.* **1958,** *58,* 925.
(5) Hearle, J.W.S. *Polymers and Their Properties*, Vol. 1: *Fundamentals of Structure and Mechanics;* Ellis Horwood: Chichester, England, 1982.
(6) See Chapter 7 in this volume.
(7) Seymour, R. B.; Carraher, C. E., Jr. *Structure-Property Relationships in Polymers;* Plenum Press: New York, 1984.

Glues and adhesives

Question
How do glues and adhesives work?

Answer
By secondary chemical bonding.

Secondary chemical bonds, or van der Waals forces, are the forces most frequently found holding an adhesive and substrate together. Secondary bonds include London or dispersion forces, dipole–dipole interactions, and hydrogen bonds. Of these secondary bonds, the most significant for explaining adhesion are the dispersion forces, particularly for nonpolar glues such as those used in pressure-sensitive tape and rubber cement. This may seem surprising, as dispersion or van der Waals forces are so often thought of as "weak" chemical interactions. The important point to remember is that the strength of dispersion forces is highly dependent upon the distance between interacting molecules and falls off with the inverse sixth power of the distance. This fact leads to the most critical requirement for an adhesive: An adhesive must provide good wettability or intimate molecular contact between adhesive and substrate. To get good contact, adhesive and substrate should have similar surface tensions and polarities. A drop of water on a well-oiled sunbather beads up as though the water and suntan oil were trying to repel one another. This beading of the water prevents contact as much as possible between the oil and water. The water is polar with a high surface tension; the oil is nonpolar with a low surface tension. The converse of this phenomenon is also true: Entities with similar polarities and surface tensions will, when brought together, result in close molecular contact.

How strong can these "weak" London forces possibly be? For a distance of 1 nm, theoretical calculations predict a strength between adhesive and substrate to be the equivalent of 1000 atm (100 MPa) of pressure (*1*). The reason that actual joints do not show this theoretical strength is that trapped air spaces, cracks, or other defects inevitably form during the adhesion process and cause failure below the theoretical value. (A chain is only as strong as its weakest link.)

The need to be close explains why polymers with flexibility, like rubber, are better adhesives than other less flexible polymers such as polystyrene. Freedom of rotation on the molecular level may also help an adhesive snuggle up to the substrate (*2*).

Hydrogen bonding across the adhesive/substrate interface makes for good adhesion. Resorcinol-formaldehyde resins are used to adhere nylon tire cords to rubber in the making of tires (*1–3*).

$$\left[\begin{array}{c}\text{OH}\\\text{CR}_2\end{array}\right]_n \text{— Resorcinol-formaldehyde (R=H, or resorcinol link)}$$

$$\left[-\text{NH(CH}_2)_x\text{NH}-\overset{\text{O}}{\underset{\|}{\text{C}}}(\text{CH}_2)_y\overset{\text{O}}{\underset{\|}{\text{C}}}-\right]_n \quad \text{Nylon}$$

Although secondary bonding is what is most commonly found at the adhesive interface, other types of adhesion also occur: primary bonding by covalent metallic bonds forms in welding or soldering. Other mechanisms, such as diffusion, mechanical interlocking, and electrostatic attraction, also aid adhesion. A comprehensive theory of adhesion still needs to be developed.

An additional factor to consider in the strength of an adhesive is its cohesive strength. Whereas adhesive strength refers to the bond between the adhesive and substrate, *cohesive* strength refers to the strength within the adhesive itself. Improving cohesive strength is the motivation behind developing epoxies and other adhesives that polymerize and cross-link when they set.

References
(1) Kinloch, A. J. *J. Mater. Sci.* **1980**, *15*, 2141.
(2) *Handbook of Adhesives*, 2nd ed.; Skeist, I., Ed.; Van Nostrand Reinhold: New York, 1977.
(3) Petke, F. D. In *Polymer Science and Technology*, Vol. 9A; Lee, L.-H., Ed.; Plenum Press: New York, 1975; p. 177.
(4) Schneberger, G. L. *Adhesives Age* **1985**, *28*, 10.
(5) *Adhesion and Adhesives*, Vols. 1 & 2; Houwink, R.; Salomon, G., Eds.; Elsevier: Amsterdam, 1965.

Paints and surfactants

Question
Water-based paints and white household glue (for example, Elmer's Glue®) are similarly structured. What is this physical structure called and how does it work?

Answer
A latex.

A latex includes a minimum of three ingredients: water, small polymer particles, and a surfactant. The polymer is not soluble in water and is soft enough to be deformable. The surfactant stabilizes the polymer in the suspension and prevents it from precipitating.

A surfactant molecule has a polar (hydrophilic) head and a nonpolar (hydrophobic) tail. These different ends allow a surfactant to keep immiscible polar and nonpolar molecules from separating by surrounding each phase with the similar end of the surfactant molecule. In a latex, the nonpolar ends of the molecule surround the polymer particles, and the polar ends face the water (Figure 5). Spreading a latex across a substrate may be compared with depositing marbles across a surface. The water evaporates or diffuses into the substrate, and the capillary pressure of the evaporating water pulls the polymer spheres together. Because the polymer is deformable under surface tension pressures, the spheres sinter together, forming a continuous film (Figure 6). Pigments and stabilizers may be added to the latex for color, to lengthen shelf life, and for general product improvement. Polyvinylacetate polymer is used in Elmer's Glue and was used for many years in most latex paints. Because of their chemical stability, acrylics have now largely replaced other polymers in latex paints.

References
(1) Turner, G.P.A. *Introduction to Paint Chemistry*, 2nd ed.; Chapman and Hall: London, 1980.
(2) Oil and Colour Chemists' Association, Australia. *Surface Coatings, Vol. 1—Raw Materials and Their Usage;* Chapman and Hall: London, 1983.
(3) Nylen, P.; Sunderland, E. *Modern Surface Coatings;* Interscience: London, 1965; p. 665.
(4) Talen, H. W. *J. Oil & Colour Chem. Assn.* **1962**, *45*, 387.

Home Runs and humidity

Question
Why is it more difficult to hit a home run on a humid summer night than it is on a cool dry evening? (*Hint:* A baseball core is made of tightly wound wool yarn, and wool is a polyamide similar to nylon in structure [1].)

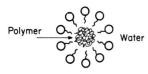

Figure 5. Action of surfactant molecules in a latex.

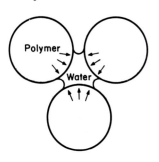

Figure 6. Action of polymer spheres as water evaporates.

Answer
Water absorption reduces elasticity.

Water hydrogen-bonds to the polar amide linkages in the wool; thus, the higher the humidity, the more water that will be absorbed into the ball. This absorbed water makes the ball heavier. The absorbed water also acts as an external plasticizer and softens the ball, causing it to lose some of its liveliness. Remember, in an elastic collision the objects that collide are not permanently deformed and kinetic energy is conserved. In an inelastic collision only momentum is conserved. The harder and less deformable the ball, the easier it is to hit a home run.

Sports Illustrated (2) reported that baking a baseball at 212 °F for 24 h causes it to lose 12 g in weight and gain 5.8% in vigor.

A polyamide such as wool or nylon forms hydrogen bonds with itself (see structure). When water is absorbed, it breaks these secondary bonds apart by forming its own hydrogen bonds with the polymer. This spreads the tangled mass of polymer chains apart, providing more room for the molecules to move around, as though partially dissolved in a solvent. This is how any external plasticizer works to make a softer, more deformable polymer. It is called an external plasticizer because it is added to the polymer rather than polymerized into the polymer when it is made. All plasticizers work as the water does in the nylon or wool example: by forming secondary bonds with a polymer, breaking the polymer-polymer secondary bonds, spreading the polymer molecules apart, and thereby giving more freedom for the chains to move. For example, polyvinyl chloride (PVC) is used to make tough, rigid, cold-water pipes. However, with a plasticizer such as dioctyl phthalate added, PVC becomes soft enough to make shower curtains.

References
(1) Rosen, S. L. *Fundamental Principles of Polymeric Materials*; John Wiley: New York, 1982; p. 85.
(2) *Sports Illustrated*; July 20, 1970; p. 22.
(3) Starkweather, H. W., Jr. In *Water in Polymers*; Rowland, S. P., Ed.; ACS Symposia Series 127; American Chemical Society: Washington, D.C., 1980; Chapter 25.

Graphite and diamonds

Question
Diamond is transparent, colorless, and hard, whereas graphite is black, slippery, and soft. What differences in bonding and structure lead to these marked differences in properties? If an organic solvent could be found that just floats graphite, would diamonds sink in it or float? (*Hint:* Diamonds form deep in the Earth from carbon that is subjected to very high pressures.)

Answer
Diamond has a tetrahedral crystal structure in which each carbon atom is linked to four others by strong covalent bonds. The strength of the diamond comes from this sp^3-hybrid bonding. The crystal structure is isometric, meaning that it extends equally in all directions. A diamond is really an enormous single molecule! The lowest excited states are so high in energy that light absorption is far in the ultraviolet, so that pure diamonds are clear and colorless.

The crystal structure of graphite consists of sheets of carbon atoms. Each carbon atom is covalently connected by sp^2-hybrid bonds to three others in the sheet, forming connecting hexagons (like a chicken wire fence); the sheets are held together only by van der Waals forces, so they slide easily over one another. The many low-lying π^* excited states of the p_z orbitals absorb light so effectively that the substance is black.

Figure 7. Crystal structures of (a) diamond and (b) graphite.

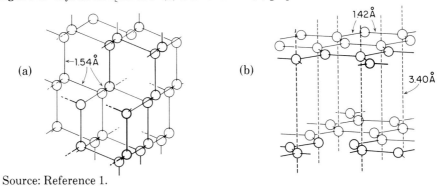

Source: Reference 1.

The van der Waals forces in graphite are much weaker than the covalent bonds in diamond. In fact, they are so much weaker that on the Mohs scale (a scale used by mineralogists to rate hardness) graphite is considered a 1 or 2 and diamond is a 10, the hardest known substance.

As a general rule, the hardness of a mineral increases with increasing packing density. The packing index, a measure of packing density, is equal to the (volume of ions/volume of unit cell) × 10. The packing index of graphite is much lower than that of diamond. Note the distances between carbon atoms illustrated in Figure 7. This is reflected in the density of the two crystals. Diamond has a density of 3.50 g/cm^3, but graphite has only 2.09–2.23. Therefore, if an organic solvent were found that just floated graphite, diamonds would sink. Diamond, formed under high pressure deep in the Earth, has a greater density and is the thermodynamically more stable form of carbon at high pressure.

References
(1) Berry, L. G.; Mason, B. *Mineralogy;* W. H. Freeman: San Francisco, 1959; p. 175.
(2) Orlov, Y. L. *The Mineralogy of the Diamond;* John Wiley: New York, 1977.
(3) *Diamonds, Myth, Magic, and Reality;* Legrand, J., Ed.; Crown Publishers: New York, 1980.
(4) Keith, F. *Modern Mineralogy;* Prentice Hall: Englewood Cliffs, N.J., 1974.
(5) Zoltai, T.; Stout, J. H. *Mineralogy;* Burgess: Minneapolis, 1984.

Antique window glass

Question
In the restored colonial village of Williamsburg, Virginia, the original windowpanes may be identified by measuring the thickness of the panes at the top and bottom. Antique window glass is thicker at the bottom than the top. What causes this phenomenon?

Answer
Engineers who study materials say that the physical behavior of glass can best be described as that of a liquid with a very high viscosity. A unique feature of glass is that when it melts, there is no sharp transition between solid and liquid as there is for ice and water. Heated glass will slowly soften and gradually liquefy. It is this softening property that makes it possible to blow glass.

Soda-lime glass, which is the most common windowpane glass, has a composition $Na_2O \cdot CaO \cdot 6SiO_2$. Each silicon atom bonds tetrahedrally to four oxygen atoms, and each oxygen bonds to two silicon atoms, resulting in a random three-dimensional network of silica. The Na_2O and CaO occupy spaces available within the loose silica structure, causing occasional breaks in the network.

When glass is heated, the increased motion of the molecules causes some of the silicon-oxygen bonds to break, and the various chains of the polysilicate structure become free to move. If enough bonds break, the glass will flow. At room temperature, the process of breaking and remaking Si—O bonds still occurs, but the equilibrium of the Si—O bond-breaking reaction lies far to the left and is an extremely slow process.

$$Si-O \rightleftarrows Si + O$$

However, under the continued stress of gravity, window glass will flow like a liquid—albeit very slowly. The viscosity of glass is estimated to be between 10^{16} and 10^{20} Poise. (1 Poise = 1 gm·cm^{-1}·s^{-1}. Water at room temperature has a viscosity of 1×10^{-2} Poise.) After several hundred years, the effects of this high-viscosity flow become noticeable on a macroscopic level.

The engineering term for this phenomenon is *creep*. Creep is an important property to study in order to determine the durability of a material. Objects made of polymers, particularly long-chain polymers without much cross-linking, are especially susceptible to creep deformations, as an applied stress can cause the chains to slip by one another, deforming the shape of the object.

References
(1) Doremus, R. H. *Glass Science;* John Wiley & Sons: New York, 1973; Chapters 3 and 6.
(2) Flinn, R. A.; Trojan, P. K. *Engineering Materials and Their Applications*, 2nd ed.; Houghton Mifflin: Boston, 1981; Chapter 7.
(3) Brady, G. S. *Materials Handbook*, 10th ed.; McGraw-Hill: New York, 1971; p. 360.

Platinum and hydrogen

Question

If a stream of hydrogen gas is directed onto finely divided platinum, the metal will glow white hot as long as the hydrogen stream is maintained. This can be repeated any number of times. What could be happening?

Answer

The platinum acts as a catalyst, greatly accelerating the rate of reaction between H_2 and O_2 of the air to form H_2O as the product. The very exothermic reaction occurs so rapidly it heats the metal to incandescence. The platinum acts by forming Pt—H and Pt—O bonds on the metal surface; the surface species then react rapidly to form product water. Powdered platinum works best because of its high surface area per unit weight. The discovery of the reaction by Doebereiner in 1821 led to the first lighters in 1823. Berzelius coined the word *catalysis* in 1835 to describe the discovery.

Reference
   (1) Collins, P.M.D. *Platinum Metals Rev.* **1986**, *30*, 141.

Shape selectivity in zeolites

If toluene, methyl chloride, and a little of the Lewis acid $AlCl_3$ are mixed together at about 80 °C, the major product is *meta*-xylene. If the reaction takes place at 0 °C, the products are primarily *ortho*- and *para*-xylene. At high temperatures, the *ortho*- and *para*-xylene species actually form (as at the lower temperatures) but quickly isomerize to *m*-xylene, since it is the product favored at equilibrium. This thermodynamic favoring of *m*-xylene makes it difficult to produce high yields of *p*-xylene, the product that is most desired because it can be oxidized to terephthalic acid and then reacted with ethylene glycol to form the polyester Dacron.

Industrially, *p*-xylene is made from a mixture of toluene and methanol with a zeolite catalyst. Zeolites are a class of solid acid catalysts made from porous crystalline silica and aluminum oxide. These heterogeneous catalysts are sometimes impregnated with platinum, calcium, iron, and other cations to enhance or change reactivity.

A computer-drawn picture of one face of a mordenite crystal shows the honeycomb structure characteristic of zeolites.

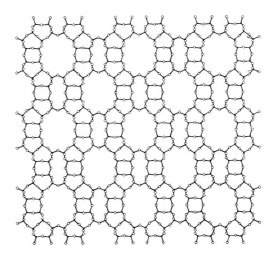

Question

How is it possible to get up to 97% *p*-xylene with zeolite catalysts when thermodynamic considerations clearly favor *m*-xylene formation? In addition, raising the temperature with these catalysts increases the *p*-xylene selectivity! Explain.

Answer

The secret of the zeolite's success is that it may be synthesized to have pores of a discrete size with most of the active sites on the inside of the pore structure. Although xylenes form inside the pores in equilibrium proportions, only the *p*-xylene is able to diffuse out easily. In one type of zeolite catalyst, the diffusion coefficient for *p*-xylene is 1000 times as large as those for the other isomers. To maintain equilibrium conditions, more *p*-xylene is formed within the pores as it diffuses out. Raising the temperature increases the rate of diffusion, causing the rate of *p*-xylene appearance to increase also.

Para-xylene selectivity is just one example of the use of shape-selective zeolite catalysts. Another example is hydrocarbon formation from methanol over Mobil's HZSM-5 catalyst. Methanol over a zeolite catalyst reacts to form hydrocarbon chains. The length of the chain is controlled by the size of the pore. A long and tangled chain would have difficulty diffusing out of a small pore. Therefore, Mobil's remarkable catalyst only produces molecules with fewer than 10 carbons. This predominantly C8 product is, for all intents and purposes, gasoline!

References
Xylene Synthesis with Zeolites
(1) Csicsery, S. M. *Chemistry in Britain* **1985**, *21*(5), 473.
Zeolites—General
(2) *Zeolite Chemistry and Catalysis;* Rabo, J. A., Ed.; ACS Monograph 171; American Chemical Society: Washington, D.C., 1976.
(3) Smith, K. W.; Starr, W. C.; Chen, N. Y. *Oil Gas J.* **1980**, *78*(21), 75.
(4) Weisz, P. B. *Pure Appl. Chem.* **1980**, *52*, 2091.
(5) Dwyer, J.; Dyer, A. *Chem. Ind.;* April 2, **1984**; p. 237.

Substrate selectivity in zeolites

Question
Treatment of microporous zeolite particles with a solution of $RhCl_3$ gives a heterogeneous catalyst that can be used to hydrogenate olefins under mild conditions. There is, however, little or no substrate selectivity; e.g., cyclopentene and 4-methylcyclohexene add H_2 at the same rate. Treatment of the zeolite catalyst with PBu_3 greatly enhances the selectivity for cyclopentene, whereas PMe_3 kills all activity. How could this be possible?

Answer
Treatment with $RhCl_3$ exchanges Rh(III) ions for the Na^+ ions originally in the zeolite. With the untreated catalyst, most of the hydrogenation occurs with the readily accessible surface Rh, which displays no selectivity. PBu_3 poisons the surface Rh, but is too bulky to enter the pores of the zeolite, which are typically 5–8 Å in diameter—large enough to allow the olefins to enter and favoring the smaller cyclopentene. PMe_3 is small enough to enter and poison all of the Rh sites (*1*). In addition, the amount of H_2O coating the walls of the channels can slow the movement of methyl cyclohexene relative to cyclopentene; selectivity for the smaller olefin can be as large as 50:1 (*2*).

References
(1) Huang, T.-N.; Schwartz, J. *J. Am. Chem. Soc.* **1982**, *104*, 5244.
(2) Corbin, D. R.; Seidel, W. C.; Abrams, L.; Herron, N.; Stucky, G. D.; Tolman, C. A. *Inorg. Chem.* **1985**, *24*, 1800.
(3) For a review of zeolites and their uses, see Schwochow, F.; Puppe, L. *Angew Chem. Int. Ed. Engl.* **1975**, *14*, 620.

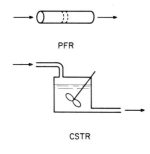

Figure 8. Schematic representations of two types of continuous flow reactor vessels. (a) Plug flow reactor, (b) continuous stirred tank reactor.

Types of continuous reactors

Two types of continuous flow reactor vessels are commonly used in industry for producing high-volume chemicals. One is the plug flow reactor (PFR) and the other the continuous stirred tank reactor (CSTR), both shown schematically in Figure 8.

In an ideal PFR, reactants travel in narrow bands or plugs, with mixing within a plug but no mixing backward or forward; the time after mixing the reactants is proportional to the distance traveled down the tube. In an ideal CSTR, the liquid is so well stirred that the reactor contents are uniform throughout, and the exit stream composition is the same as the composition in the reactor.

Question
For a series of first-order reactions $A \xrightarrow{k_1} B \xrightarrow{k_2} C$ in which B is the desired product, the yield of B can be significantly higher from a PFR than from a CSTR. Explain. If $k_2 = 10k_1$ and the holdup time τ (the reactor volume divided by the liquid flow rate) is such that $k_1\tau = 0.05$, what is the percent conversion of A in each type of reactor? What is the percent yield of B in each case?

Answer
A PFR behaves much as does the batch reactor more familiar to chemistry students, in which the reactants are added at once and the reaction is allowed to proceed with time, except that in a PFR time is measured as distance down the reactor divided by the linear flow rate. As in a batch reactor, the concentration of B increases with time (or with distance travelled down the reactor). Therefore, since the rate of formation of C is dependent upon the concentration of B, significant amounts of C will not begin to form until B builds up. On the other hand, with a CSTR the entire reactor contains the high concentration of B found in the exit stream. Since the average residence time spent in

each reactor is identical, more C will form in the CSTR than in the PFR because, on the average, [B] is higher in the CSTR, giving more opportunity for C to form.

To study this problem on a quantitative basis, a material balance for component A is made for each reactor.

1. A material balance for A in the PFR can be made for each thin plug of liquid flowing through the reactor.

n_A(in) $- n_A$(out) = amount of A that has reacted
$= -k_1$ [A]·vol

n_A(in) is the flow of A (in mol/s) into the plug and vol is its volume. By letting the volume of the plug go to zero we find

$d[A]/dt = -k_1[A]$

This is just the equation for a first-order chemical reaction with rate constant k_1, except that t is now the time spent traveling down the reactor. (The concentrations at a fixed point along the tube do not change with time.) Similarly,

$d[B]/dt = k_1[A] - k_2[B]$
$d[C]/dt = k_2[B]$

The solution (1) if [A] = [A$_0$] at $t = 0$ while
[B] and [C] = 0 at $t = 0$, is

$[A] = [A_0]e^{-k_1 t} = [A_0]e^{-0.05} = 0.9512 [A_0]$

$[B] = \dfrac{[A_0]k_1}{k_2 - k_1}(e^{-k_1 t} - e^{-k_2 t}) = [A_0]/9(e^{-0.05} - e^{-0.5}) = 0.0383[A_0]$

% Conv of A = 100([A$_0$] − [A])/[A$_0$]
= 100(1.0 − 0.09512) = 4.88% in the PFR

% Yield of B = 100[B]/([A$_0$] − [A])
= 100(0.0383)/(1.0 − 0.09512) = 78.5%

2. A material balance for A in the CSTR can be carried out over the whole reactor, as the concentration of A is constant throughout.

(Flow rate) ([A$_0$] − [A]) = k_1[A] · vol

Remembering that τ = vol/flow

[A$_0$] − [A] = $-k_1\tau$[A]
[A] = [A$_0$]/(1 + $k_1\tau$) = [A$_0$]/1.05 = 0.9523[A$_0$]

Similarly,

[B] = $k_1\tau$[A]/(1 + $k_2\tau$) = $k_1\tau$[A$_0$]/(1 + $k_1\tau$)(1 + $k_2\tau$)
= 0.0317 [A$_0$]

% Conv of A = 100 (1.0 − 0.9523) = 4.76% in the CSTR

% Yield of B = 100 (0.0317)/(1.0 − 0.9523) = 66.7%.

Note that while the percent conversion of A in the CSTR is slightly lower (the concentration of A is only 0.9523 [A$_0$] throughout the CSTR), the yield of B is substantially less. The excellent mixing of a CSTR (desirable if reactants have to be mixed before reaction can occur) with the yield advantages of a PFR can be approached by staging a number of CSTRs in series (2).

References
(1) Frost, A. A.; Pearson, R. G. *Kinetics and Mechanism;* John Wiley and Sons: New York, 1953; p. 153.
(2) Hill, C. G., Jr. *An Introduction to Chemical Engineering Kinetics and Reactor Design;* John Wiley and Sons: New York, 1977.
(3) Smith, J. M. *Chemical Engineering Kinetics;* McGraw-Hill: New York, 1981; Chapter 3.

Electric eels

Question
An electric eel, such as a Torpedo, can produce a terrific shock, up to 600 V, whereas a common flashlight battery produces only about 1.5 V. How is the eel able to do it? Could a victim be shocked in salt water, a very good electrical conductor?

Answer
Electric fish and eels possess electric organs specifically developed to produce high voltages. These organs are made from modified muscle cells in which the contractile material is absent and only the outside membrane remains. Each flattened electrical cell, called an electroplate, has one nerve running to it. Like ordinary nerve and muscle cell pairs, the nerve stimulates the cell and causes an electric potential across a membrane.

The membrane of the electroplate nearest the nerve ending is excitable, which means that its electrical potential is able to change. At rest, this excitable electroplate membrane, called the innervated face, has a resting potential of −90 mV. The membrane on the other side of the flattened cell is called the noninnervated face and is a highly folded, nonexcitable membrane. At rest, the noninnervated membrane also has a resting potential of −90 mV.

When the membrane is excited, acetylcholine, a neurotransmitter, interacts with receptor sites on the membrane by opening channels for sodium and potassium ions to pass through. At rest, the inside of the cell has a low concentration of sodium ions and a high concentration of potassium ions. Outside the cell the reverse is true, with a high sodium ion concentration and low potassium. The inside versus outside concentration difference is greater for sodium than potassium. When the Na^+/K^+ channels open up, the substantial difference in concentrations between the inside and the outside causes large amounts of sodium ions to rush into the cell and a smaller quantity of potassium ions to flow out. The driving force is just ordinary diffusion caused by the tendency of a chemical species to move from an area of high concentration to an area of low concentration. Sodium can be said to be moving down its concentration gradient.

The net inward flow of charge changes the potential across the excitable membrane from −90 mV to +60 mV, giving an overall potential of 150 mV between the outside faces of the innervated (+60 mV) and noninnervated (−90 mV) membranes (see Figure 8).

Figure 9. Electroplates at rest (a) and when excited (b).

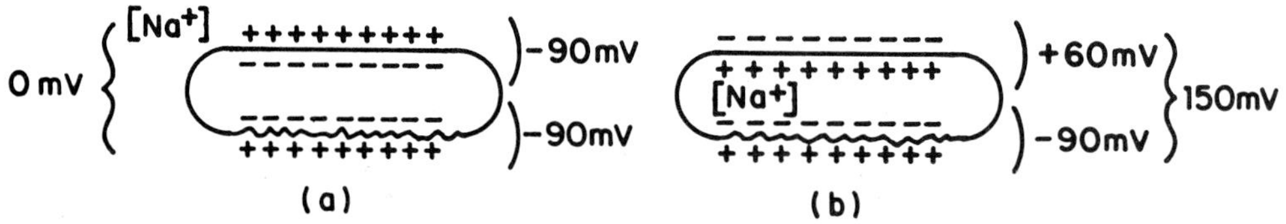

The electroplates in an electric organ are placed in series so that the voltages add together. Therefore, an electric organ with 4000 electrocells is capable of producing a voltage of 600 V. This high a voltage can easily give a stunning shock to humans or fish in the water. Zoological researchers report that handling these animals can be quite painful. Electric eels use their unique ability to both protect and feed themselves. They hover over their choice of dinner, flip their electrical switch, and zap! The food is ready to be eaten.

Salt water is such a good electrical conductor that electric rays in the sea (with large areas of electroplates in their wings) have evolved the ability to generate currents between their wings of up to 30 amps for short periods of time! Their voltage is not as high as that of freshwater eels—only 200–300 V. It's a wonder the things don't electrocute themselves.

References
(1) Grzimek, B. *Grzimek's Animal Life Encyclopedia*, Vol. 4, Fishes I; Van Nostrand Reinhold: New York, 1973; p. 121.
(2) Bennet, M.V.L. In *Fish Physiology*, Vol. 5; Hoar, W. S.; Randall, D. J., Eds.; Academic Press: New York, 1971.
(3) Stryer, L. *Biochemistry*, 2nd ed.; W. H. Freeman: San Francisco, 1981; pp. 887, 889.

Chemical oscillations and biorhythms

Question
A small but growing set of chemical systems contains *chemical oscillators*, such that in the overall conversion of reactants to products there are one or more intermediates whose concentrations undergo periodic oscillations in time (and sometimes also in space). How is this possible?

Although periodic behavior in chemical systems has been known for only a relatively short time (the first reported purely chemical system was in 1921 [1]), the coupling of physical and chemical effects, especially in biological systems, gives a large number of periodic phenomena, some of which have been known for millennia. How many can you name? (*Hint:* Two examples are heartbeat and menstruation.)

Answer
Three conditions seem to be necessary for chemical oscillation:
1. The system should be far from equilibrium.
2. There should be feedback; i.e., a species produced should have an effect on the rate of its own formation.

3. The chemical system should exhibit bistability; i.e., under a given set of initial conditions it should be able to exist in two different steady states (2).

A simple mathematical model to describe oscillatory behavior of the populations of game (G) and predators (P) using three equations was described by Lotka (3, 4).

$$G \rightarrow 2G \tag{1}$$
$$P + G \rightarrow 2P \tag{2}$$
$$P \rightarrow D \tag{3}$$

In the absence of predation (2), the game population increases by (1). Increase of predator population in (2) depends on the presence of game. Finally, in (3), predators eventually die (old age).

Chemical systems usually require a greater number of differential equations. The Belousov-Zhabotinskii reaction, involving the overall oxidation of malonic acid ($HOCOCH_2CO_2H$) by bromate ions (BrO_3^-) to carbon dioxide, formic acid (HCO_2H), and bromomalonic acid ($HOCOCHBrCO_2H$) in the presence of H^+, Br^-, and Ce^{+3}, has been simulated on a computer using 18 elementary steps and 21 chemical species (2). It can give chemical waves in three dimensions, including toroidal scrolls (5).

If coupling of physical and chemical effects can occur (for example, if O_2 has to be transferred from the gas phase to the liquid to react [6]), the possibilities for oscillatory behavior become rich. Not surprisingly, there are dozens of examples of oscillatory phenomena in living systems.

Some periodic biological phenomena and their typical time periods are listed below.

Phenomenon	Typical period
Nerve firing	0.001 s
EEG pulses	0.1 s
Heartbeat	1 s
Firefly flashes	3 s
Breathing	4 s
Hiccups	20 s
Sleep stage	90 min
Eating	6 h
Sleeping/waking	24 h
Menstruation of women	1 mo
Hibernation of bears	1 yr
Migration of geese	1 yr
Spawning of salmon	3 yr
Human birth/procreation	25 yr
Population cycles of wolves and moose	38 yr

In some cases, such as sleeping/waking and the migration of geese, the phenomena are clearly coupled to sunlight through the variation of length of day with season. (Some people suffer severe depression during winter because of the effects of the length of the day on their biochemistry; their condition can be cured by artificially lengthening the period of daylight with fluorescent lights [7]). The 38-year cycle refers to an isolated population of wolves and moose in Isle Royale National Park in Lake Superior (8).

> Our bodies are constantly influenced by cycles of day and night, seasons, tides, magnetic and gravitational forces, and, some claim, even such environmental forces as lunar phases. The human brain responds to these cycles by mirroring them with cycles of its own: sleep and waking, the ebb and flow of hormones, periodicities in our levels of alertness and competence. Even our emotions follow cycles. . . . Neuroscientists are currently exploring how all these "biological clocks" are constituted, what winds them up, and how they run down (10).

References

(1) Bray, W. C. *J. Am. Chem. Soc.* **1921**, *43*, 1262.
(2) Epstein, I. R.; Kustin, K.; DeKepper, P.; Orban, M. *Sci. Am.* March **1983**, 112.
(3) Lotka, A. J. *J. Am. Chem. Soc.* **1920**, *42*, 1595.
(4) Higgins, J. "The Theory of Oscillating Reactions," Applied Kinetics and Chemical Reaction Engineering; ACS. Publications: Washington, D.C., 1967; pp. 143–86.
(5) Welsch, B. J.; Gotmatam, J.; Burgess, A. E. *Nature* **1983**, *304*, 612.
(6) Roelofs, M. G.; Wasserman, E.; Jensen, J. H.; Nader, A. E. *J. Am. Chem. Soc.* **1983**, *105*, 6329.
(7) Arendt, J. *New Scientist;* July 5, **1985**, 36.

(8) Peterson, R. O.; Page, R. E.; Dodge, K. M. *Science* **1984,** 1350.
(9) Field, R. J. *Am. Sci.* **1985,** *73*, 142.
(10) Restak, R. M. *The Brain;* Bantam Books: New York, 1984; p. 3.

General References

(1) Brown, F. A., Jr.; Hastings, J. W.; Palmer, J. D. *The Biological Clock;* Academic Press: New York, 1970.
(2) Sweeney, B. M. *Rhythmic Phenomena in Plants;* Academic Press: New York, 1969.
(3) Bunning, E. *The Physiological Clock;* Springer-Verlag: New York, 1967.
(4) *Biological Aspects of Circadian Rhythms;* Mills, J. N., Ed.; Plenum Press: New York, 1973.
(5) Moore-Ede, M. C.; Sulzman, F. M.; Fuller, C. A. *The Clocks That Time Us;* Harvard University Press: Cambridge, Mass., 1982.
(6) Kuffler, S. W.; Nicholls, J. G.; Martin, A. R. *From Neuron to Brain;* Sinauer Associates: Sunderland, Mass., 1984.
(7) Luce, G. G. *Biological Rhythms in Psychiatry and Medicine;* U.S. Department of Health, Education, and Welfare, Public Health Service Publication No. 2088, 1970.

Epilogue

We would like to close our chapter with a poem, which perhaps best captures the feelings of awe and wonder that some courses in physical chemistry may be able to impart.

What Kind of Chemistry

> What kind of chemistry can respond to a smile?
> Or open a green bud?
> Or make a meadowlark sing?
> What kind of chemistry is there in an embrace?
> Or in a song?
> Or in a sunset that sets the sky on fire?
> What kind of chemistry is there in your lover's eyes?
> Or in the smell of new hot baked bread?
> Or in the blueness of mountain sky?
> What kind of chemistry?

C.A.T.
9/28/79

Notes added in proof

For section on candles burning in zero gravity (pp. 16–17)
Some nice photographs showing the effects of burning a candle in 0, 1, or 2g gravity or in an electric field can be found in Carleton, F. B.; Weinberg, F. J. *Nature* **1987,** *330*, 635.

For section on substrate selectivity in zeolites
Although linear hydrocarbons normally are preferentially oxidized at the secondary (weaker) C–H bonds along the chain in preference to the primary C–H bonds in the terminal methyl groups, we have been able to reverse the selectivity using O_2 as the oxidant in properly designed zeolite catalysts—thus mimicking the behavior of the ω-hydroxylase enzymes with an inorganic system.
Herron, N.; Tolman, C. A. *J. Am. Chem. Soc.* **1987,** *109*, 2837.
Herron, N.; Tolman, C. A. *Am. Chem. Soc., Div. Petr. Chem.* **1987,** *32* (1), 200.

For section chemical oscillations and biorhythms
In 1751 Linnaeus designed a flower clock for the 12 hours between 6 a.m. and 6 p.m. based on the characteristic times of opening and closing of the leaves and petals of 12 flowering plant species. It works even on cloudy days because of the built-in biological clocks. (Gen. Ref. 5, p. 12)

Chapter 4
CONTINUITY OF SPECIES IN PHYSICAL AND CHEMICAL PROCESSES

PETER R. RONY
Department of Chemical Engineering
Virginia Polytechnic Institute and State University
Blacksburg, Va. 24061

Slattery (1) and others (2) have pointed out that a common philosophy is the basis for fluid mechanics, thermodynamics, heat transfer, and mass transfer. Students in a wide range of disciplines benefit from a study of an integrated approach to such subjects. In addition to gaining an understanding of the approach, they develop a set of skills that permits them to represent the important aspects of physical problems in mathematical terms.

Fluid mechanics, heat transfer, and mass transfer are commonly taught in chemical engineering departments in a year-long undergraduate or graduate sequence entitled *transport phenomena*, the successor to the traditional *unit operations* sequence that was taught prior to (and many years after) the publication of Bird, Stewart, and Lightfoot's classic text in 1960 (2). Though all three subjects are potentially valuable to a chemistry student, there is not time to teach them in the typical undergraduate chemistry curriculum. But this time limitation does not mean that they should be totally excluded, either.

Of the three subjects, mass transfer is by far the most important to a practicing chemist, who encounters the subject regularly in chemical kinetics, chemical reactors, chemical separations (e.g., chromatography and distillation), diffusion-controlled reactions, and electrochemistry (3). A wonderful opportunity exists to tie these subjects into an efficiently taught, unifying whole that can be comprehended by juniors in a physical chemistry course. The basis for this unification is the conservation-of-species equation (also called the continuity-of-species equation), which has been used for more than half a century in the modeling of macroscopic chemical systems in which simultaneous transport, reaction, and equilibrium occur. A case can, in fact, be made that conservation of species ranks as one among five fundamental theoretical approaches to the chemical sciences that physical chemistry students must master (the other four being statistical mechanics, kinetic theory, quantum mechanics, and thermodynamics).

Continuity of species is an approach that breaks down disciplinary barriers. A summary passage from the article "Diffusion and Chemical Transformation: An Interdisciplinary Excursion" by Paul B. Weisz (4) is appropriate here:

Phenomena versus Disciplines: A Postscript

> We have traced the significance of the competition between molecular reaction and molecular diffusion, as expressed by the Φ criterion, across many disciplines. It is interesting to note how researchers in many branches of the sciences have struggled with problems that arise from a basically common phenomenon, one that does not 'belong' to any special discipline. We need to consider how we might best prepare ourselves and our students to understand and to deal with basic phenomena of general applicability. While we continue a healthy trend to erase interdisciplinary boundaries, we must remember that any collection of items as complex as those that constitute human knowledge can only be sensibly stored, managed, or propagated with some unifying structure. Can we find structures that are orderly, basic, and interdisciplinary? Can the teaching of human experience and knowledge be organized around *phenomena*? The diffusion-transformation interaction is but one example of a general phenomenon; there are many others. For example, energy...feedback...zero

...kinetic interaction of assemblies...each represents a general concept or phenomenon that has meaning, implications, and applicability across an enormous sector of experiences and disciplines, physical, biological, social, physiological, medical, psychological, and others. Furthermore, each can be experienced, enjoyed, and understood in some form at nearly every stage of educational development.

Edward de Bono made a somewhat similar point in an article in CHEMTECH (5):

Up til quite recently, the world was full of *things*. That is to say, the most useful and respected way to look at the world was in terms of static *things*. These *things* were given very definite names and fitted into great schemes of categories and classifications and sub-classifications. Every *thing* had its proper place and, being a static *thing*, it stayed there.

It is often felt that *processes* are much more complicated than *things*. I do not believe this is so at all. It is true that you can see cars and bicycles and boiled eggs every day, but rarely see *polarization* or *feedback* or *pattern*. But this may only be because one has never learned to look for them. It can be just as much fun to look for *feedback* in daily life as it is to look for snails in the garden or buttercups on a country wall. And one can become just as quick at recognition.

De Bono's first list of process concepts included the following items (5):

to feedback	to manage	to forecast
to optimize	to control	to permit
to catalyze	to transfer	to convert
to sample	to oscillate	to filter
to polarize	to trigger off	to extrapolate
to amplify	to regenerate	to communicate

These two quotes convey the spirit of my Chemical Microengineering course (an elective course that was taught for the first time in 1973), which, to a large degree, is based upon the continuity-of-species equation. *Chemical microengineering* is defined as the study of the engineering of small chemical systems. Another part of the spirit of this course is captured by Tomas Hirschfeld in his article, "Providing Innovative System Monitoring and Reliability Assessment Through Microengineering" (6):

Mainly, however, microengineering beyond the electronics domain gives us the possibility of designing objects in an entirely new fashion since the significance of the physical and chemical principles involved changes as the scale of the system decreases. In an attempt to learn from experience, we have kept in mind that the biological world has been designing on the micrometer scale for eons and, thus, furnishes us with some impressive examples of microengineering....

We have undertaken a systematic study of the physics, chemistry, and engineering of the microscopic domain. To do this, we have had to bring together a number of fields that initially do not appear to have much in common. The obvious starting point was microelectronics and its fabrication technology. However, unless we were to build everything out of silicon and only in two dimensions, we could not stop there. We soon realized that microscopic sizes are the normal design range of the biological world, and, after a billion years or so, nature has accumulated considerable expertise in this area. Through a careful search of the molecular biology and entomology literature, we identified a number of useful biological devices and design principles that can be incorporated in our microengineered devices.

This leads directly to the first major technical task in microengineering—identifying the scaling laws. The scaling laws describe the variation in the relative importance of different phenomena as things change in size. For example, weight increases with the cube of a linear dimension but strength increases only with the square....

Objectives of this chapter

The objectives of this chapter are:

1. To provide a unifying structure for teaching the continuity-of-species equation to undergraduate chemists.

2. To discuss, briefly, several key phenomena, including diffusion; convection; diffusion-controlled reaction; convection-controlled reaction (continuous reactor); equilibrium-controlled separation; and membrane permeation.

3. To illustrate the use of the continuity-of-species equation in one-dimensional systems.

4. To discuss the concept of a *linear multistate chemical system* and to demonstrate its use for several of the following applications: chromatography; diffusion-controlled reaction in a porous catalyst; tubular catalytic reaction; carrier diffusion; phase-transfer catalysis; ideal distillation; multiphase catalysis; diffusion; multiphase catalytic reaction; ideal extraction; and chemical kinetics.

Phenomena and processes

In the spirit of the comments of Weisz and de Bono, what are the phenomena and processes associated with applications of the continuity-of-species equation? First are the processes, which can be expressed as verbs:

to diffuse	to distribute
to convect	to partition
(to undergo convection)	to separate
to react	to catalyze
to accumulate	to undergo a diffusion-controlled reaction
to equilibrate	to undergo a convection-controlled reaction

Second are the phenomena, which can be expressed as nouns:

diffusion	partitioning
convection	separation
reaction (chemical kinetics)	(diffusion-controlled convection)
accumulation	catalysis
equilibration	diffusion-controlled reaction
distribution	convection-controlled reaction

Third are the parameters that serve to measure the amount or magnitude associated with a given process or phenomenon:

diffusion coefficient	distribution coefficient	volume fraction
velocity	partition coefficient	area-to-volume ratio
rate constant	volume	rate
equilibrium constant	interfacial area	volumetric rate
space time	surface area	surface rate

Fourth, and finally, is a list of techniques based upon the application of the continuity-of-species equation, as well as devices and equipment in which the phenomena and processes occur:

distillation	carrier transport	extractor
extraction	reactor	diffusion cell
chromatography	chromatograph	heterogeneous catalyst
parametric pumping	distillation column	fixed-bed reactor
field-flow fractionation		

We do not have time to discuss all such concepts, but will demonstrate an intellectual framework in which their discussion can occur naturally in a physical chemistry course.

Definitions

Practitioners in the chemical sciences frequently work with multicomponent systems. Definitions are in order (*1, 7*).

body of species i   A set, the elements of which are called particles of species i or material particles of species i (*1*).

chemical system	A specified region, or portion of matter, containing a definite amount of one or more species in one or more phases.
phase	A region of space in a chemical system that is separated from the remainder by a clearly definable surface and within which the thermodynamic properties differ from the remainder when the whole system is in thermal equilibrium.

There are three distinct types of phases—the *gas* phase, the *liquid* phase, and the *solid* phase—which correspond to the three fundamental states of matter.

interface	A surface that forms the boundary between two phases.
homogeneous system	A chemical system that contains only a single phase.
heterogeneous system	A chemical system that contains two or more phases and at least one type of interface.
mixture	A heterogeneous or homogeneous aggregation of different substances. The components of a mixture may be separated by simple mechanical or physical means, in sharp contrast to a chemical compound, which cannot be so separated.
substance	A term that is used to designate a pure chemical compound or a mixture of such compounds.
component	One of the distinct molecular, ionic, atomic, or aggregative species composing a mixture.

A component must be distinguished from

constituent	In thermodynamics, one among the smallest number of chemical substances that need to be specified in order to reproduce a given chemical system.

In this chapter, components will be denoted by the subscript i, which has been selected because it is a most appropriate subscript for FORTRAN programs that describe multicomponent chemical systems. Phases will be denoted by the subscript s. Thus, a component i present in phase s will be denoted by the double subscript is. I believe that it is important for students to recognize early the widespread existence of multiphase, multicomponent systems.

In the context of the engineering term *rate process*, the word rate refers to the ratio of change in a physical quantity to a change in time or distance. Usually, a function of time is implied whenever rate is used without qualification. The definitions for rate and rate process, as they are used in chemical engineering, can be stated as follows:

rate	The ratio of the differential change in a physical quantity to a differential change in time or distance. When no qualification is stated, a differential change in time is implied (8).
rate process	A continuous or discrete action or series of actions or events that can be expressed as a rate or as the sum or difference of individual rates (8).

What is a continuity-of-species equation?

Conservation Laws

In attempting to deal with many different types of rate processes, scientists and engineers have used one of the outstanding ideas of science, the *conservation law*, to great advantage. A conservation law states that a "conserved" physical quantity—mass, energy, momentum, or charge—can be neither created nor destroyed by ordinary physical or chemical means. Such a law can be written in terms of rates, with each rate reflecting the rate process that contributes to the conservation of the physical quantity in question. It is important, therefore, to identify for a given physical situation the types of rate processes that occur; to express such rate processes by appropriate rate expressions; and to employ the rate expressions in the appropriate conservation law equations. In doing all this, scientists and engineers engage in a form of bookkeeping in which they account for the influence that each rate process has upon a physical or chemical system (8).

Mass Balance Equation for an Individual Species

Slattery (*1*) treats fluid mechanics, thermodynamics, heat transfer, and mass transfer in terms of six fundamental postulates. In this chapter, we shall focus only on his sixth postulate, "The time rate of change of mass of each species i (i = 1, 2, 3, ...N) in a multicomponent mixture is equal to the rate at which the mass of species i is produced by homogeneous chemical reactions" (*1*, page 447), which is called the *mass balance equation for an individual species*, the *continuity-of-species equation*, or the *conservation-of-species equation*. This postulate, and the equations that result from it, unifies many areas of chemistry, biochemistry, and chemical engineering. As was mentioned previously, the equation plays an important organizing role that undergraduate students in physical chemistry should find valuable.

Accumulation, Transport, and Reaction

The continuity-of-species equation is discussed in chemical engineering courses on transport phenomena, mass transfer, separations, and reactor engineering. The field of transport phenomena is the study of fluid flow, heat transfer, and mass transfer or, in other words, the study of the conservation of momentum, energy, and mass. Mass transfer is the engineering subject concerned with the accumulation, transport, and reaction of molecules—of any size, complexity, purity, or state of aggregation—as governed by the continuity-of-species equation and other conservation laws. By *accumulation* we mean whether or not the number of molecules of a specific type increases, decreases, or remains constant in a given volume (the *control volume*) of a physical system under consideration. By *transport* we mean the movement of molecules across the boundaries (the *control surface*) of the control volume. By *reaction* we mean the conversion of a specific molecule (or specific molecules) into some other type of molecule (or molecules) within the control volume (*8*).

The Continuity-of-Species Equation

For a specific molecule under consideration, the continuity-of-species equation can be stated in word form as follows:

Rate of accumulation of molecule in control volume	=	Rate of transport of molecule into control volume	−	Rate of transport of molecule out of control volume
		+ Rate of production of molecule in control volume	−	Rate of loss of molecule in control volume

The rate of production and loss terms include the interconversion of one molecule into another. In this word equation, there are two transport terms—across the control surface *into* and *out of* the control volume—and two reaction terms—*production* and *loss* within the control volume. This equation is valid only for a control volume and associated control surface that are both precisely specified (*8*).

The Conserved Physical Quantity

A rate was defined as the ratio of the differential change in a physical quantity to a differential change in time. What is the physical quantity associated with the continuity-of-species equation? It is some measure of the amount of the molecule under consideration. For example, the amount of sucrose (the molecule) in a cup of coffee (the chemical system) can be written in terms of grams, kilograms, pound-mass, number of molecules, parts per million, mole fraction, mass fraction, and so forth. The fact that the amount of a substance can be expressed in so many different ways is one of the persistent problems in the field of mass transfer.

Let us select the *number of molecules* of a substance as the physical quantity that we use in the continuity-of-species equation. Of course, this quantity is extremely large in most systems, as one mole of a substance contains 6.023×10^{23} molecules. The rate of change in the continuity-of-species equation, therefore, has units of *molecules/second*. If the volume of the system is measured in units of cubic meters, then the *density* or *concentration* of the molecules—the number of molecules per unit volume—has units of *molecules/m^3*.

In summary, at least three physical quantities or ratios are important in a system to which the continuity-of-species equation is applied:

- rate (molecules/time)
- density or concentration (molecules/volume)
- total number (molecules)

Two other commonly encountered quantities are
- flux (molecules/area-time)
- volumetric rate of change (molecules/volume-time)

The product of a flux times an area yields a rate; the product of a volumetric rate times a volume also yields a rate (8).

The continuity-of-species equation: A unifying theory in physical chemistry

Just what or how does the continuity-of-species equation unify? Figures 1 and 2 provide one answer. In the next section, the continuity-of-species equation contains rate terms associated with the following phenomena: diffusion, convection, and reaction. Not immediately evident from the equation is a fourth phenomenon, equilibration, that becomes important in multiphase systems and is treated by the *linear multistate chemical systems* approach later in this chapter. In terms of these phenomena, the continuity-of-species equation can be stated in word form for an appropriately chosen control volume and associated control surface:

Net rate of   = Net rate of + Net rate of + Net rate of
accumulation   diffusion   convection   reaction

Figures 1 and 2 subdivide the applications of the equation into subsets that are based upon simpler forms of the equation: diffusion only; convection only; reaction only; diffusion and convection only; diffusion and reaction only; convection and reaction only; and diffusion, convection, and reaction. Figure 1 applies to steady-state systems (see next section), and Figure 2 applies to non-steady-state systems (see next section). With the aid of these subsets, the basic relationship among diverse aspects of the chemical sciences—equilibrium separations, chromatography, carrier transport through a membrane, diffusion-controlled reaction in a catalyst pellet, tubular reactors, batch reactors, residence time distribution studies, multiphase catalysis, chemical kinetics, multiphase catalytic reactors, diffusion, reaction chromatography, and parametric pumping—becomes clear: *All are manifestations of a single conservation law, the continuity of species.* This chapter will elaborate on this simple, but important, message.

Figure 1. Subsets of the steady-state continuity-of-species equation. Steady-state reactors and separation systems are commonplace in industry. Steady-state diffusion-controlled systems can be found in biological systems and catalyst pellets, for example.

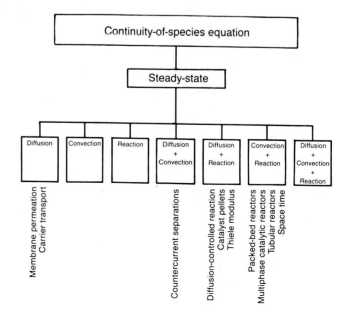

Figure 2. Subsets of the non-steady-state continuity-of-species equation. Techniques such as chromatography, parametric pumping, field-flow fractionation, reaction chromatography, and multiphase catalysis fall into this overall category.

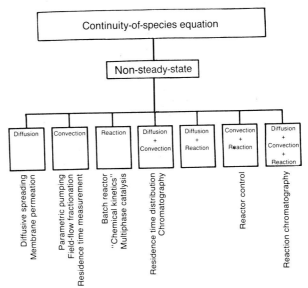

The continuity-of-species equation for a homogeneous chemical system

One-Dimensional Equation

Students in physical chemistry should find a one-dimensional continuity-of-species equation (with constant coefficients) (*1, 2, 9*) for species i to be very useful when they consider the macroscopic behavior of dynamic chemical systems. The final one-dimensional equation

$$\underbrace{\frac{\partial c_{is}}{\partial t}}_{\text{accumulation}} \underbrace{- D_{is}\frac{\partial^2 c_{is}}{\partial z^2}}_{\text{diffusion}} + \underbrace{v_s \frac{\partial c_{is}}{\partial z}}_{\text{convection}} = \underbrace{R_{is}}_{\text{reaction}} \qquad (1)$$

is applicable for both Cartesian coordinates and cylindrical coordinates. Even though the derivation is not given here, it is important to remember the assumption that concentration gradients for species i in the remaining two coordinate directions are zero. This assumption permits the three-dimensional equation (Equation 2) to be simplified. Equation 1 is a powerful equation that applies to a variety of one-dimensional homogeneous chemical systems. In this equation,

c_{is} = concentration of species i in phase s (molecules/volume)
v_s = mass-average velocity of phase s (length/time)
D_{is} = diffusion coefficient of species i in phase s (length^2/time)
R_{is} = net molar rate of production or loss of species i in phase s (moles/volume-time)
z = z-coordinate direction in Cartesian or cylindrical coordinate geometry (length)
t = time

The symbol ∂ designates a partial derivative.

Three-Dimensional Equation

The corresponding three-dimensional continuity-of-species equation for species i is

$$\underbrace{\frac{\partial c_{is}}{\partial t}}_{\text{accumulation}} \underbrace{- D_{is}\nabla^2 c_{is}}_{\text{diffusion}} + \underbrace{v_s \nabla c_{is}}_{\text{convection}} = \underbrace{R_{is}}_{\text{reaction}} \qquad (2)$$

where the symbol ∇ designates the del vector operator (*1, 2, 9*). Equation 1 states that the concentration of species i in a phase s of defined control volume and control surface changes as a result of the convection of species i across the control surface, the diffusion of species i across the control surface, and reactions that produce and destroy species i within the control volume.

Special Cases of the Continuity-of-Species Equation for a Homogeneous Chemical System

Equation 1 is a partial differential equation that has two independent variables: time and the z-coordinate distance. In many cases, convection, diffusion, and reaction do not occur simultaneously in the chemical system under consideration, so the continuity-of-species equation can be simplified to produce equations that are subsets of either Equation 1 or Equation 2. Seven special cases of Equation 1 can be identified (7); they correspond to the boxes in Figure 2.

Case 1. Diffusion only

$$\frac{\partial c_{is}}{\partial t} - D_{is}\frac{\partial^2 c_{is}}{\partial z^2} = 0 \qquad (3)$$

Case 2. Convection only

$$\frac{\partial c_{is}}{\partial t} + v_s\frac{\partial c_{is}}{\partial z} = 0 \qquad (4)$$

Case 3. Reaction only

$$\frac{\partial c_{is}}{\partial t} = R_{is} \qquad (5)$$

Case 4. Diffusion and convection only

$$\frac{\partial c_{is}}{\partial t} - D_{is}\frac{\partial^2 c_{is}}{\partial z^2} + v_s\frac{\partial c_{is}}{\partial z} = 0 \qquad (6)$$

Case 5. Diffusion and reaction only

$$\frac{\partial c_{is}}{\partial t} - D_{is}\frac{\partial^2 c_{is}}{\partial z^2} = R_{is} \qquad (7)$$

Case 6. Convection and reaction only

$$\frac{\partial c_{is}}{\partial t} + v_s\frac{\partial c_{is}}{\partial z} = R_{is} \qquad (8)$$

Case 7. Diffusion, convection, and reaction

$$\frac{\partial c_{is}}{\partial t} - D_{is}\frac{\partial^2 c_{is}}{\partial z^2} + v_s\frac{\partial c_{is}}{\partial z} = R_{is} \qquad (9)$$

An eighth case corresponds to thermodynamic equilibrium in the system.

Case 8. No diffusion, convection, or reaction (thermodynamic equilibrium)

$$[\text{No continuity-of-species equation can be written}] \qquad (10)$$

Steady-State vs. Non-Steady-State Systems

The *IEEE Standard Dictionary of Electrical and Electronics Terms* (*10*) defines the term *steady state* in the following manner:

steady state
1. That in which some specified characteristic of a condition, such as value, rate, periodicity, or amplitude, exhibits only negligible change over an arbitrarily long interval of time. 2. The condition of a specified variable at a time when no transients are present.

The term *transient* means changing with time. The one-dimensional equation (1) specifies that the concentration of the species i depends both on distance and time, $c_{is}(z,t)$. Such a system is called *non-steady-state* because the variable c_{is} is dependent upon time. The corresponding equation for a steady-state one-dimensional system is

$$\underbrace{-D_{is}\frac{d^2 c_{is}}{dz^2}}_{\text{diffusion}} + \underbrace{v_s\frac{dc_{is}}{dz}}_{\text{convection}} = \underbrace{R_{is}}_{\text{reaction}} \qquad (11)$$

Observe that the time derivative, $\partial c_{is}/\partial t$, is now equal to zero; the dependent variable $c_{is}(z)$ depends only on distance; and the spatial derivative is now an ordinary derivative rather than a partial derivative.

Steady state does not mean equilibrium. The condition of steady state means that the competing rate processes—diffusion, convection, and reaction—that comprise the continuity-of-species equation are in such perfect balance that the magnitude of the concentration at any point in the system does not change with time.

What Is a Flux?
The *International Dictionary of Physics and Electronics* (*11*) defines *flux* as follows:
flux 1. A quantity proportional to the surface integral of the normal (perpendicular) force field intensity over a given area,

$$\text{Flux} = K \int_s \vec{F}_N \cdot d\vec{S}$$

where $\vec{F}_N$ is the normal component of a field (e.g., gravitational, electric, magnetic), and K is the constant of proportionality between the field and the flux density (permitivity, permeability, etc.). 2. A term which denotes the volume or mass of fluid or particles transferred across a given area perpendicular to the direction of flow in a given time.

The *Dictionary* continues (*11*):
There are many specific applications in physics of the term flux. For electromagnetic radiation, it signifies the energy per unit time, or the power passing through a surface. For photons or particles, flux is the number per unit time passing through a surface. In nuclear physics, flux commonly means the product nv, where n is the number of particles per unit volume and v is their mean velocity.

In the context of the continuity-of-species equation, *flux* can be defined as the number of molecules transferred in a given increment of time across a given area perpendicular to the direction of flow. Frequently the given area is the control surface associated with a control volume. Molecules are transferred by the transport processes of diffusion and convection.

What Is a Convection Process?
Convection can be defined as the process by which matter is transported from one point in space to another as a result of streaming fluid motions. Other than the velocity of the fluid stream, the quantity of greatest interest is the *convective mass transfer flux*, which is the quantity of a substance that is transferred, in a given increment of time, by convection across a given area perpendicular to the direction of flow. The convective flux can be expressed as moles per unit area per unit time, molecules per unit area per unit time, or mass per unit area per unit time. As an example,

$$\text{Convective molar flux of species } i = c_{is} v_s \text{ mol/area-time} \tag{12}$$

where v_s is the local *mass-average velocity* (*1, 2*).

What Is a Diffusion Process?
Diffusion can be defined as the process by which matter is transported from one point in space to another as a result of random molecular motions. The quantity of greatest interest is the *diffusive mass transfer flux*, which is the quantity of substance that is transferred, in a given increment of time, by diffusion across a given area perpendicular to the direction of the diffusive flow. The diffusive flux can be expressed as moles per unit area per unit time, molecules per unit area per unit time, or mass per unit area per unit time. As an example,

$$\text{Diffusive molar flux of species } i = -c_s D_{is} \nabla x_{is} \text{ mol/area-time} \tag{13}$$

This equation is known as *Fick's first law of diffusion*, in which the proportionality factor is the *diffusion coefficient*, D_{is} (*1, 2, 7, 9*). c_s is the total molar concentration of phase s, and x_{is} is the mole fraction of species i in phase s.

What Is a Reaction Process?
The term *reaction*, taken in its broadest sense, means any change in the internal constitution of a chemical system. Such a change may be in the physical state of the species or may be the result of chemical transformation among the molecules in the system. The change in the physical state of the species is called a *physical reaction*, whereas a chemical transformation among the molecules is called a *chemical reaction*. A chemical reaction, in the Dalton sense of the term, is an elementary reaction that can be represented by a stoichiometric equation, where the stoichiometric coefficients are usually integers whose magnitudes are less than 20. A

physical reaction is a reaction which, when all of the constituents are included, generally cannot be represented by a stoichiometric equation. For example, the dissolution of a gas in a liquid usually does not proceed in a stoichiometric manner, as one mole of a gas does not dissolve for every m moles of solvent, where m is an integer.

According to the law of definite proportions, the increase in moles of a species i that is being formed in a chemical reaction is proportional to its stoichiometric coefficient ν_i in the reaction. The following equation can therefore be written (12),

$$n_{is} - n_{is}^0 = \nu_i \xi \tag{14}$$

where ξ is called the De Donder *extent of reaction* or simply the *reaction coordinate*, and n_{is}^0 is the initial number of moles of species i in phase s ($t = 0$).

The rate of reaction at time t, defined as

$$\text{Rate of reaction} = \frac{d\xi}{dt} \tag{15}$$

is "the ratio of the change (positive, zero, or negative) in the extent of reaction to the time interval dt (always taken as positive) during which the change takes place" (12). The rate of reaction can also be expressed in terms of the rate of production or consumption of an amount of a substance, expressed in moles per unit time, molecules per unit time, or mass per unit time. As an example,

$$\frac{dn_{is}}{dt} = \nu_i \frac{d\xi}{dt} \text{ mol/time} \tag{16}$$

The rate of reaction can also be expressed in terms of the rate of production or consumption of a quantity of substance per unit control volume, expressed in moles per unit time per unit volume, molecules per unit time per unit volume, or mass per unit time per unit volume. As an example,

$$R_{is} = \frac{dc_{is}}{dt} = \frac{1}{V_s} \frac{dn_{is}}{dt} = \frac{\nu_i}{V_s} \frac{d\xi}{dt} \text{ mol/volume-time} \tag{17}$$

Observe that the volumetric rate of reaction, R_{is}, has units of amount per unit volume per unit time.

The rate of reaction is generally a function of temperature, composition, and pressure. The form of the rate function is a central problem in applied chemical kinetics. Once the function is known, information can frequently be inferred concerning the rates of individual reaction steps. Also, a reactor can be designed for carrying out the reaction under "optimum" conditions (13, 14).

Initial Conditions

The solution of a differential equation involves a process of integration. As in any integration process, undefined constants appear. One constant is associated with the first-order time derivative in Equation 1, and two constants are associated with the second-order spatial derivative. The former constant is obtained through the use of a single *initial condition*, and the latter two constants are obtained through the use of two *boundary conditions*.

Initial conditions for continuity-of-species equations are frequently concentration profiles within the control volume,

$$c_{is}(z,t)|_{t=0} = c_{is}(z,0) = c_{is}(z) \tag{18}$$

A limiting case of $c_{is}(z)$ is a constant profile,

$$c_{is}(z,t)|_{t=0} = c_{is}^0 \tag{19}$$

Three examples of initial conditions are illustrated in Figure 3: uniform initial concentration c_{is}^0 in a homogeneous chemical system that has no spatial concentration gradients; a concentrated amount of substance (represented by the hypothetical Dirac delta function) at point $z = 0$ in a homogeneous system; and uniform initial concentration c_{is} in a plane polymer membrane of thickness $2L$.

Boundary Conditions; Symmetry

Boundary conditions for continuity-of-species equations are more diverse than initial conditions. If a diffusion process is present, two boundary conditions are required for each coordinate direction—for example, the concentration of species i at two

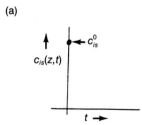
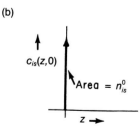
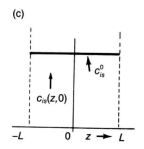

Figure 3. Three examples of initial conditions: (a) uniform initial concentration with no spatial gradients, (b) a pulse of a substance, and (c) uniformly distributed initial concentration.

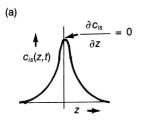
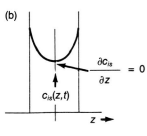
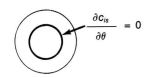

Figure 4. Three examples of symmetry boundary conditions: (a) peak maximum, (b) middle of permeable membrane or catalytic membrane, and (c) ring in disk chromatography.

different boundaries, the flux of species i at two different boundaries, or the concentration and flux of species i at a single boundary. The symmetry of the physical system frequently leads to at least one simple boundary condition.

Symmetry boundary conditions are commonly encountered in systems in which diffusion is an important process. The symmetry condition takes the form

$$\frac{\partial c_{is}}{\partial z} = 0 \tag{20}$$

which is a shorthand for the boundary condition statement that the diffusive flux is zero,

$$-D_{is}\frac{\partial c_{is}}{\partial z} = 0 \tag{21}$$

a condition that occurs when equal concentrations of species i exist on either side of a real or hypothetical boundary. Examples, shown in Figure 4, include (a) the center of a diffusively spreading amount of species i in a homogeneous chemical system; (b) the center of a plane polymer membrane that is exposed on both sides to permeant i, which permeates throughout the membrane by diffusion; and (c) any point on a single solute ring in circular paper chromatography. Observe that in all three cases the boundary is hypothetical.

Other typical boundary conditions are shown in Figure 5: (a) the flux of i is zero at a single boundary; (b) the flux of i is a constant (not zero) at a single system boundary; (c) the concentration of i is zero at positive and/or negative infinity; (d) the flux of i is zero at the hypothetical boundaries of positive and/or negative infinity; (e) the concentration of species i is zero at two different boundaries; and (f) the concentration of i is constant at the surface of a polymeric plane membrane. One of the most interesting boundaries in chemistry is the electrode boundary in electrochemical systems; see Reference 3 for further details.

Figure 5. Six examples of boundary conditions: (a) flux is zero, (b) flux is constant, (c) concentration is zero at $+/-$ infinity, (d) flux is zero at $+/-$ infinity, (e) concentration is zero at two boundaries, and (f) concentration is constant at a surface.

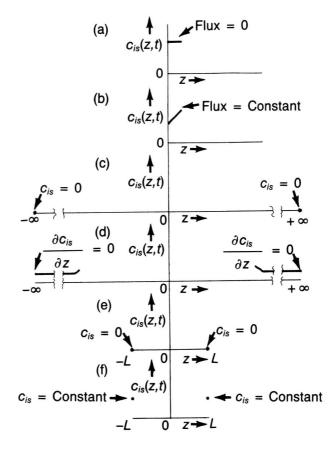

Applications: Homogeneous chemical systems

Introduction

A simple, linearized one-dimensional continuity-of-species equation for a chemical species i within a homogeneous phase s was given in the preceding section,

$$\underbrace{\frac{\partial c_{is}}{\partial t}}_{\text{accumulation}} \underbrace{- D_{is}\frac{\partial^2 c_{is}}{\partial z^2}}_{\text{diffusion}} + \underbrace{v_s \frac{\partial c_{is}}{\partial z}}_{\text{convection}} = \underbrace{R_{is}}_{\text{reaction}} \qquad (1)$$

and is applicable for both Cartesian coordinates and cylindrical coordinates (z-coordinate direction). In this equation,

- c_{is} = concentration of species i in phase s (molecules/volume)
- v_s = mass-average velocity of phase s (length/time)
- D_{is} = diffusion coefficient of species i in phase s (length^2/time)
- R_{is} = net molar rate of production or loss of species i in phase s (moles/volume-time)
- z = z-coordinate direction in Cartesian or cylindrical coordinate geometry (length)
- t = time

As was also discussed in the preceding section, the myriad of possible solutions to Equation 1 can be grouped into seven special cases depending upon which rate processes—diffusion, convection, and reaction—are physically dominant in the chemical system under consideration. An additional set of special cases arises when one distinguishes between steady-state and non-steady-state operation. Few textbooks organize solutions to the continuity-of-species equation in this way, but such an organization is valuable to the undergraduate student who has never previously encountered the subject.

What special cases of the continuity-of-species equation occur in homogeneous chemical systems? Several come to mind immediately: the field of homogeneous chemical kinetics, already well treated in physical chemistry texts; diffusion of a solute through a homogeneous, polymeric membrane under conditions where the concentration of the solute is not high enough to influence membrane properties; diffusion of a solute within an infinite stagnant liquid; residence time distributions of a solute in a flowing liquid (provided that account is taken of mass transport in all three coordinate directions); a tubular flow reactor containing a homogeneous phase; and so forth.

If the applications of the continuity-of-species equation were limited to homogeneous systems, the equation might easily be dismissed as being of little practical importance to physical chemists. A more important message, however, is that *the range of applications of the continuity-of-species equation is substantially more varied and of much greater significance in chemical systems that contain more than one phase* (or more than one type of chemical *state*, a term that will be defined in the next section). We ask the reader, therefore, to view this section as an interlude between the conceptual foundations of the continuity equation, as given in previous sections, and the exciting and far-reaching applications given in the section "Multiphase/Multistate/Coupled Chemical Systems."

Steady-State Diffusion: A Method To Continuously Transfer Molecules across a Membrane

Consider a steady-state chemical system that consists of a thin, plane polymeric membrane that separates two compartments of well-mixed liquid (Figure 6). In one compartment, the concentration of a permeating species i is c_{i1}^0; in the other compartment, the concentration of i is zero. In order to maintain such steady-state conditions, flow systems are required, one to replenish lost i from the top compartment and the other to remove permeated i from the bottom compartment. Such conditions can be achieved experimentally and can simplify the boundary conditions (and thus the mathematics) of the continuity-of-species model. It is assumed that neither reaction nor convection occurs within the polymer membrane, which is considered to be the control volume. The two surfaces of the membrane are the control surfaces, and the membrane is assumed to be infinitely long.

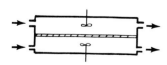

Figure 6. Steady-state diffusion system: plane membrane separating two well-mixed fluid compartments.

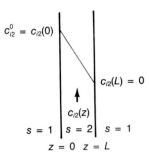

Figure 7. Linear concentration profile in the plane membrane that separates two well-mixed fluid compartments. See Equation 25.

A schematic diagram of the solution to the theoretical model is shown in Figure 7. The defining steady-state ordinary differential equation for the control volume is simply

$$-D_{i2}\frac{d^2 c_{i2}}{dz^2} = 0 \tag{22}$$

Equation 22 is an *ordinary differential equation* because of the steady-state assumption for the system; the time derivative is absent, and hence partial derivatives are not required.

The boundary conditions are established at the $z = 0$ plane boundary and the $z = L$ plane boundary:

B.C. 1: $\quad c_{i2}(z,t) = c_{i2}^0 \quad$ at $z = 0$ \hfill (23)

B.C. 2: $\quad c_{i2}(L) = 0 \quad$ at $z = L$ \hfill (24)

The solution of Equation 22 subject to boundary conditions 23 and 24 is

$$c_{i2} = c_{i2}^0 \left(1 - \frac{z}{L}\right) \tag{25}$$

Equation 25 represents a straight-line concentration profile *within the membrane*.

Using Equation 13, the diffusive flux within the membrane can be calculated as follows:

$$\text{Diffusive molar flux} = -c_{i2} D_{i2} \nabla x_{i2}$$

$$= -D_{i2}\frac{dc_{i2}}{dz}$$

$$= +D_{i2}\frac{c_{i2}^0}{L} \tag{26}$$

which has units of mol/area-time. Observe how the negative sign cancels the negative slope of the concentration profile to produce a positive diffusive molar flux of species i from left to right in membrane s.

Equation 26 is somewhat deceiving, however. In order to calculate the total rate of species i that permeates through the plane membrane, Equation 26 states that one needs to know only the cross-sectional area and thickness of the membrane, the diffusion coefficient of i through the membrane, and the concentration on one surface of the membrane (assuming that the concentration on the other side is zero). What is the deception? The value of c_{i2}, which, according to the model, is a concentration within the control volume (the membrane)—on the membrane surface—and not in the external solution, which is outside of the defined control volume. It is not easy to measure the concentration of i at the surface defined by $z = 0$; thus, from an experimental point of view c_{i2}^0 remains unknown!

We have treated the membrane as if it were a homogeneous chemical system, but we have neglected the fact that the overall system is heterogeneous: It consists of a membrane and two compartments containing liquids. The concentration of i in the membrane is c_{i2}. The concentration of i in the left-hand compartment in Figure 6 is not c_{i2}^0, but rather c_{i1}^0. Note that subscript $i2$ refers to species i in the membrane, whereas subscript $i1$ refers to i in the top solution compartment. If the concentration of i is small, a good approximation to the *jump boundary condition* at the solution/membrane interface is a linear *partition coefficient*,

$$\kappa_{i2} = \frac{c_{i2}^0}{c_{i1}^0} \text{ m}^3 \text{ solution/m}^3 \text{ membrane} \tag{27}$$

which is not dimensionless, but rather has dimensions of mol/vol membrane divided by mol/vol solution. It is a common error in heterogeneous chemical systems for scientists and engineers to overlook the fact that the volume units associated with different phases refer to different physical entities and cannot be cancelled.

The correct form of Equation 26, referred to a measurable quantity such as c_{i1}^0, is, therefore,

$$\text{Diffusive flux} = \kappa_{i2} D_{i2} \frac{c_{i1}^0}{L} \text{ mol/m}^2\text{-s} \tag{28}$$

The permeation rate, in mol/s, through the membrane is

$$\text{Permeation rate} = \frac{\kappa_{i2}D_{i2}A_{\perp 2}c_{i1}^0}{L} \text{ mol/s} \tag{29}$$

an algebraic equation that is convenient for back-of-the-envelope calculations. The quantity $A_{\perp 2}$ is the cross-sectional area of the membrane. The *permeation coefficient* for a permeating solute through a membrane is defined as the product of the solubility (or partition coefficient) and the diffusivity,

$$\text{Permeation coefficient} = \kappa_{i2}D_{i2} \; \frac{m^2}{s} \text{-m}^3 \text{ solution/m}^3 \text{ membrane} \tag{30}$$

Equation 29 can therefore be viewed as the product of a permeation coefficient, a geometrical factor, and a driving force:

Permeation rate = (permeation coefficient) (geometrical factor) (driving force)

$$= \kappa_{i2}D_{i2} \cdot \frac{A_{\perp 2}}{L} \cdot (c_{i1}^0 - 0)$$

Equation 29 provides a first estimate of permeation rates through polymeric membranes such as dried paint, Saran wrap, polymer films on semiconductor films, and so forth. It states the rule that *the rate of permeation of a species i through a plane membrane is inversely proportional to the thickness, and directly proportional to the product of the diffusion coefficient of i within the membrane, the solubility of i in the membrane, and the external surface area.* For example, a small atom such as He has a larger diffusion coefficient in polymers than organic molecules, but fails to permeate rapidly because it is relatively insoluble in polymers; this principle was the basis for the Llewelyn membrane separator used in Varian GC/MS systems 15 years ago. An important assumption that has been made is that the diffusing species is present in very small concentrations, so that it does not change the properties of the membrane for permeating solutes. A brief table of diffusivities and solubilities in solids is given in Geankoplis (*15*).

One can further complicate this analysis of the diffusion of a solute i through the plane membrane by considering the state of the liquid films immediately adjacent to the membrane in both compartments. If the compartments are not "well stirred" in some manner, or if the permeation rate through the membrane is very high, a boundary layer resistance to molecular diffusion develops in the liquid films. We do not treat this topic in this chapter because it represents a separate issue in the field of mass transfer and requires some knowledge of fluid mechanics for its conceptual and theoretical treatment. Boundary layer diffusion is of fundamental importance to chemistry, biochemistry, and chemical engineering, but *is a topic that should be treated only after a student develops a reasonable understanding of the characteristics of the continuity-of-species equation for different types of chemical systems.* Boundary layer mass transfer is not an essential concept for understanding the fundamental principles of homogeneous or heterogeneous catalysis, chromatography, chemical reactors, carrier-mediated diffusion, or equilibrium separations.

Non-Steady-State Diffusion: Diffusion in an Infinite Stagnant Liquid

Consider a non-steady-state homogeneous chemical system that consists of a solution in which a known quantity of species i is rapidly introduced as a thin, plane source. For example, assume that i is in the form of a thin, readily soluble membrane that is rapidly inserted into the solution at $t = 0+$ without creating any disturbance in the liquid (Figure 8). We must go to such extremes of hypothetical experimental technique in order to keep the problem one-dimensional.

Initially, the concentration of i in solution is assumed to be zero. Also, it is assumed that there is no convection or reaction in the chemical system: The only rate process is the diffusion of i, from a concentrated plane source initially, to the remainder of the liquid volume, which is assumed to be infinite in both the positive and negative z-coordinate directions.

The defining non-steady-state partial differential equation is

$$\frac{\partial c_{is}}{\partial t} - D_{i2}\frac{\partial^2 c_{is}}{\partial z^2} = 0 \tag{31}$$

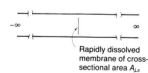

Figure 8. Thin, readily soluble membrane that is rapidly inserted into a one-dimensional liquid chamber.

which has a symmetry boundary condition at $z = 0$,

$$\text{B.C. 1:} \quad \frac{\partial c_{is}}{\partial z} = 0 \quad \text{at } z = 0 \tag{32}$$

a flux condition at both positive and negative infinity,

$$\text{B.C. 2:} \quad \frac{\partial c_{is}}{\partial z} = 0 \quad \text{at } z = \infty \text{ and } z = -\infty \tag{33}$$

and an initial condition that assumes that i is concentrated at $z = 0$ and is not present anywhere else in the solution,

$$\text{I.C. 1:} \quad c_{is}(z,0) = 0 \quad \text{for } z \neq 0 \text{ and}$$

$$2n_{is}^0 = A_{\perp s} \int_{-\infty}^{\infty} c_{is}^0(z)dz = A_{\perp s} c_{is}^0 \int_{-\infty}^{\infty} \delta(z)dz \tag{34}$$

where $A_{\perp s}$ is the cross-sectional area of the system and $\delta(z)$ is the Dirac delta function, a mathematical function and convenient physical fiction. In other words, the total amount of i in the system at any time is $2n_{is}^0$ mol, where the factor of 2 is selected for mathematical convenience.

The solution to Equation 31 subject to the one initial and two boundary conditions is

$$C_{is}(z,t) = \frac{n_{is}^0}{A_{\perp s}\sqrt{\pi D_{is} t}} \exp\left[-\frac{z^2}{4 D_{is} t}\right] \tag{35}$$

This result is obtained mathematically by noting the symmetry of the system: The behavior for values of z greater than zero is the mirror image of the behavior for values of z less than zero. Equation 35 is an expression for a Gaussian peak. Perhaps the easiest way to arrive at such a solution is to use the symmetry characteristic of the system and Laplace transforms, which are applicable only for linear differential equations (16–19). For further details about non-steady-state diffusion, see the classic text by Crank, *The Mathematics of Diffusion* (20).

Figure 9 illustrates the concentration profile behavior for species i at different times. The peak shapes resemble those observed in chromatography, only in this case there is no convection, so the centroid for i remains stationary at $z = 0$.

Figure 9. Concentration profiles, as a function of time, for non-steady-state diffusion into an infinite stagnant liquid. See Equation 35.

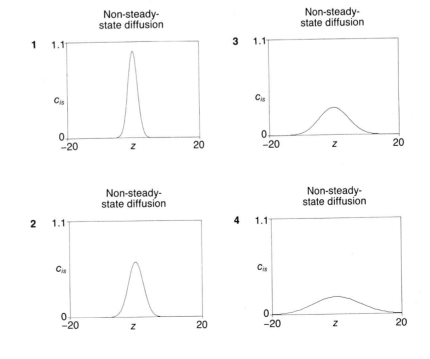

Non-Steady-State Diffusion: Diffusion into a Semi-Infinite Solid

While on the subject of non-steady-state diffusion, it is appropriate to mention two other problems that are commonly encountered in chemical systems: (1) non-steady-state diffusion into a thick (semi-infinite) solid, and (2) non-steady-state diffusion into a thin solid such as a polymeric membrane. Solutions to such problems are standard in textbooks on the mathematics of partial differential equations (*17, 18*) and transport phenomena (*2, 21*).

Consider a solid that is so thick that it can be considered to be semi-infinite, which means that it extends from $z = 0$ to $z = \infty$ (Figure 10). The surface of the slab is at $z = 0$, and diffusion proceeds into the slab. The concentration at the boundary, $z = 0$, is assumed to be c_{i2}^0 *within the solid* at its surface.

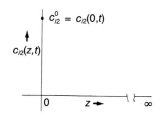

Figure 10. Initial condition for diffusion into a semi-infinite solid.

The non-steady-state partial differential equation, boundary conditions, and initial condition are

$$\frac{\partial c_{i2}}{\partial t} - D_{i2} \frac{\partial^2 c_{i2}}{\partial z^2} = 0 \tag{36}$$

$$\text{B.C. 1:} \quad c_{i2}(0,t) = c_{i2}^0 \quad \text{at } z = 0 \tag{37}$$

$$\text{B.C. 2:} \quad \frac{\partial c_{i2}}{\partial z} = 0 \quad \text{at } z = \infty \tag{38}$$

$$\text{I.C. 1:} \quad c_{i2}(z,0) = 0 \quad \text{at } t = 0 \quad \text{for } z > 0 \tag{39}$$

The solution to Equation 36 is well known,

$$c_{i2}(z,t) = c_{i2}^0 \, \text{erfc}\left[\frac{z}{2\sqrt{D_{i2}t}}\right] \tag{40}$$

where erfc is the complementary error function. Figure 11 provides examples of concentration profiles within the solid at increasing times.

Figure 11. Concentration profiles, as a function of time, for non-steady-state diffusion into a semi-infinite solid. See Equation 40.

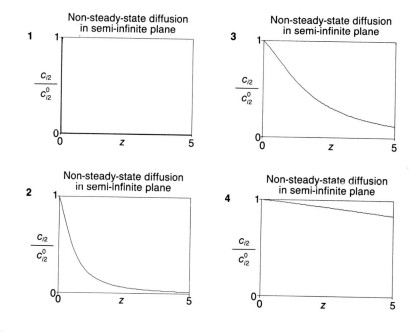

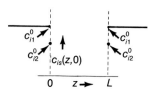

Figure 12. Boundary and initial conditions for non-steady-state diffusion into a planar membrane. The external liquid is assumed to be infinite and well mixed.

Non-Steady-State Diffusion: Diffusion into a Thin Membrane

A third common homogeneous chemical system that involves only non-steady-state diffusion is shown in Figure 12. A membrane-soluble species *i* permeates on both sides of a thin, plane polymeric membrane that is L in thickness. The non-steady-state differential equation is the same as Equations 31 and 36, only the boundary conditions are different.

$$\frac{\partial c_{i2}}{\partial t} - D_{i2} \frac{\partial^2 c_{i2}}{\partial z^2} = 0 \tag{41}$$

B.C. 1: $c_{i2}(L,t) = c_{i2}^0$ at $z = L$ (42)

B.C. 2: $c_{i2}(0,t) = c_{i2}^0$ at $z = 0$ (43)

I.C. 1: $c_{i2}(z,0) = 0$ at $t = 0$ for $0 < z < L$ (44)

The solution, again standard, is

$$\frac{c_{i2}^0 - c_{i2}}{c_{i2}^0} = \frac{4}{\pi} \sum_{n=1}^{\infty} \frac{1}{n} \sin\left(\frac{n\pi z}{L}\right) \exp\left(-\frac{(n\pi)^2 D_{i2} t}{L^2}\right) \quad n = 1, 3, 5, \ldots \quad (45)$$

Figure 13 illustrates the sequence of concentration profiles that would be observed experimentally, provided that the proper instrumentation were available. In general, concentration profiles are rarely measured; instead, the membrane is weighed as a function of time and the diffusion coefficient and solubility determined by fitting the data to a model in which the concentration profile is integrated over the entire membrane. Alternatively, the amount of uptake of solute i from the two compartments is measured as a function of time and fit to a suitable integrated model.

Figure 13. Concentration profiles, as a function of time, for non-steady-state diffusion into a planar membrane of thickness L. See Equation 45.

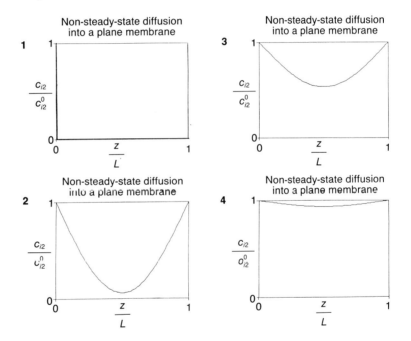

Non-Steady-State Convection: Residence Time in an Ideal Flowing System
This is a simple example of the application of the continuity-of-species equation. Assume a flowing fluid stream. A small amount of solute i is injected at point $z = 0$ and detected downstream at point $z = L$ (Figure 14). If the time between injection and detection is measured, the stream velocity can be calculated.

The defining non-steady-state partial differential equation appears more complex than it really is:

$$\frac{\partial c_{is}}{\partial t} + v_s \frac{\partial c_{is}}{\partial z} = 0 \quad (46)$$

Figure 14. Non-steady-state convection in a tubular flow system.

A simple, but powerful, transformation can be used to simplify Equation 46. Define a new parameter u_s as

$$u_s = z - \int_0^t v_s dt \tag{47}$$

$$du_s = dz - v_s dt \tag{48}$$

By employing the definition for the total differential of c_{is},

$$dc_{is} = \left(\frac{\partial c_{is}}{\partial u_s}\right)_t du_s + \left(\frac{\partial c_{is}}{\partial t}\right)_{u_s} dt \tag{49}$$

we obtain

$$\left(\frac{\partial c_{is}}{\partial t}\right)_z = -v_s \left(\frac{\partial c_{is}}{\partial u_s}\right)_t + \left(\frac{\partial c_{is}}{\partial t}\right)_{u_s} \tag{50}$$

and

$$\left(\frac{\partial c_{is}}{\partial z}\right)_t = \left(\frac{\partial c_{is}}{\partial u_s}\right)_t \tag{51}$$

Substituting Equations 50 and 51 into Equation 46, we obtain

$$\left(\frac{\partial c_{is}}{\partial t}\right)_{u_s} = 0 \tag{52}$$

a result that is surprisingly simple. What is the physical significance of Equation 52? If one travels with the injected solute peak at the velocity, v_s, of the fluid stream, the peak profile does not change! The time it takes to travel from $z = 0$ to $z = L$ is called the residence time, t_R,

$$L = \int_0^{t_R} v_s dt \tag{53}$$

The velocity could be constant, which it frequently is, or it could be a function of time. The value of a constant velocity can be determined if L and t_R are both measured. See Figure 15.

Figure 15. Concentration profiles, as a function of time, for a pulse of solute as it moves through a tubular flow system in the absence of diffusion. See Equation 53.

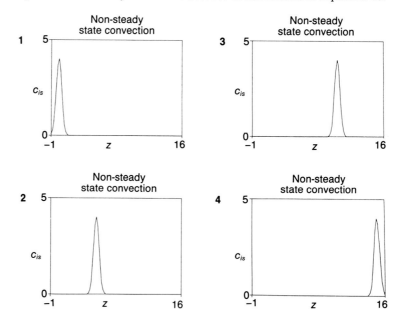

Figure 16. Non-steady-state diffusion and convection in a tubular flow system.

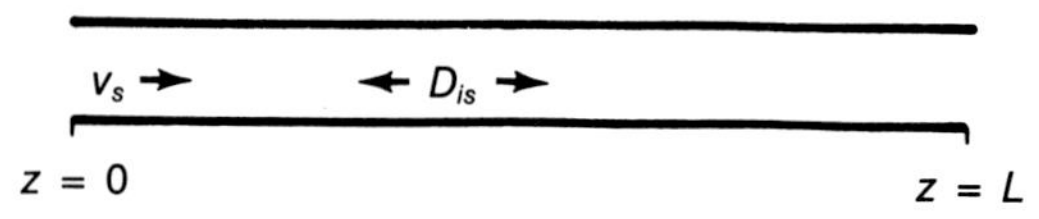

Non-Steady-State Diffusion and Convection: The Broadening of a Peak in a Flowing System

This situation is only a small extension of the last situation described. Diffusive broadening of the solute i is now permitted in the flowing stream (Figure 16). The new continuity-of-species equation becomes

$$\frac{\partial c_{is}}{\partial t} - D_{is}\frac{\partial^2 c_{is}}{\partial z^2} + v_s\frac{\partial c_{is}}{\partial z} = 0 \tag{54}$$

a more formidable equation than Equation 46. The same technique used in the previous section simplifies Equation 54 to

$$\left(\frac{\partial c_{is}}{\partial t}\right)_{u_s} - D_{is}\left(\frac{\partial^2 c_{is}}{\partial u_s^2}\right)_t = 0 \tag{55}$$

For the injection of an instantaneous pulse at $z = 0$ and $t = 0$ (that is, at $\mu_s = 0$) of $2n_{is}^0$ mol,

$$2n_{is}^0 = A_{\perp s}\int_{-\infty}^{\infty} c_{is}^0(z)dz = A_{\perp s}c_{is}^0\int_{-\infty}^{\infty}\delta(z)dz \tag{56}$$

the solution becomes,

$$c_{is}(z,t) = \frac{n_{is}^0}{A_{\perp s}\sqrt{\pi D_{is}t}}\exp\left[-\frac{(z-v_st)^2}{4D_{is}t}\right] \tag{57}$$

We now observe a combination of the behavior exhibited in Equations 35 and 52. Compare Figure 17 with Figures 9 and 15. The centroid of the steadily broadening solute peak moves at the fluid velocity v_s. The retention time of the peak centroid is given by Equation 53. Is this chromatography? Not quite, because two different solutes i and k injected at the same instant of time at $z = 0$ travel at the same velocity v_s. That is, no separation occurs. However, continuity-of-species Equation 54 and its corresponding solution, Equation 57, are only one step removed from the principle of chromatography. What is the principle? See the section on "Multiphase/Multistate/Coupled Chemical Systems."

Figure 17. Concentration profiles, as a function of time, for a diffusing solute pulse as it moves through a tubular flow system. See Equation 57.

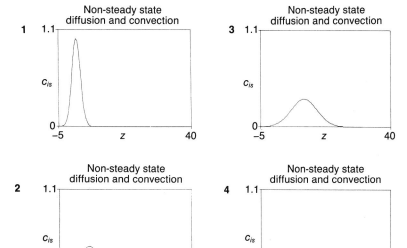

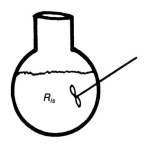

Figure 18. A constant-volume batch reactor, an example of a non-steady-state reaction system.

Non-Steady-State Reaction: The Batch Reactor

One general equation for a chemical reaction of reactant i in homogeneous phase s in a stirred batch reactor (Figure 18) is

$$\frac{\partial c_{is}}{\partial t} = R_{is} \tag{58}$$

where R_{is} is a rate expression. For further details, see Atkins (14).

One rate law that we shall use repeatedly in the section entitled "Multiphase/Multistate/Coupled Chemical Systems" is that of a pseudo-first-order chemical reaction,

$$R_{is} = -\mathrm{k}_{is}c_{is} \tag{59}$$

The solution of the continuity-of-species equation for an initial condition of

$$\text{I.C.:} \quad c_{is}(0) = c_{is}^0 \text{ at } t = 0 \tag{60}$$

is

$$c_{is} = c_{is}^0 e^{-\mathrm{k}_{is}t} \tag{61}$$

Figure 19 depicts the commonly observed profile for a first-order reaction.

Figure 19. Concentration profiles for a first-order reaction in a batch reactor (see Equation 61). The rate constant ranges from 0.02 to 5 s^{-1}.

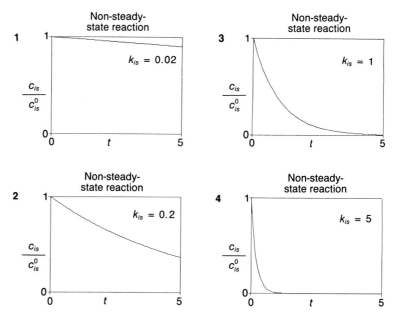

Figure 20. A cylindrical tubular flow reactor, an example of a steady-state reaction and convection system.

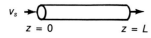

Steady-State Convection and Reaction: The Tubular Continuous Reactor and Space Time

Let us turn from non-steady-state to steady-state behavior. Consider a piece of chemical hardware called a *tubular reactor*, which is basically a metal or ceramic cylindrical pipe of appropriate diameter and length (Figure 20). Assume a flowing homogeneous phase that contains a reacting solute i. The velocity is a constant v_s, and diffusion effects are assumed to be negligible. The system is assumed to have operated for a sufficient length of time that steady-state conditions exist. The defining continuity-of-species equation is

$$v_s \frac{dc_{is}}{dz} = R_{is} \tag{62}$$

which is subject to the following boundary condition:

$$\text{B.C.:} \quad c_{is}(0) = c_{is}^0 \quad \text{at } z = 0 \tag{63}$$

An interesting transformation occurs through the definition of a *space time*,

$$\tau_s = \frac{z}{v_s}, \quad v_s \frac{dc_{is}}{dz} = \frac{dc_{is}}{d\dfrac{z}{v_s}} = \frac{dc_{is}}{d\tau_s} \tag{64}$$

which converts Equation 62 to

$$\frac{dc_{is}}{d\tau_s} = R_{is} \tag{65}$$

We have previously seen both this equation (Equation 58) and the profile (Figure 21) when the rate law is first order in reactant. The close relationship between these two equations leads to a conclusion that is well-known to chemical engineering students: A non-steady-state (batch) reactor (Figure 18) can be converted into a steady-state (continuous) reactor (Figure 20) by conducting the reaction in a flowing stream that passes through a tube of appropriate length.

Figure 21. Concentration profiles for a first-order reaction in a tubular flow reactor. See Equation 65.

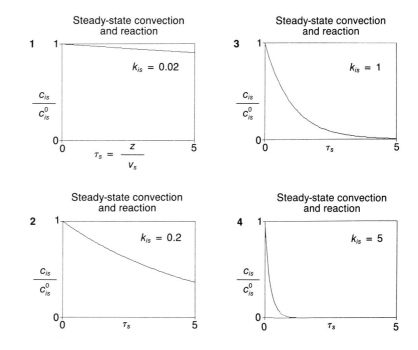

Steady-State Diffusion and Reaction: Diffusion-Controlled Reaction, Thiele Modulus, and Effectiveness Factor

One of the best written papers in the chemical microengineering literature, "Diffusion and Chemical Transformation: An Interdisciplinary Excursion," by Paul B. Weisz, has already been cited in this chapter (4). The classic example of a steady-state diffusion and reaction system is an ideal pore in a porous catalyst particle. Most catalyst pellets are cylindrical or spherical; for mathematical simplicity, we shall assume a planar pellet with a cylindrical pore (Figure 22).

Figure 22. Two types of pores in a planar, porous catalyst particle: (a) pore of length $2L$ with symmetry condition in the middle, and (b) dead-end pore of length L.

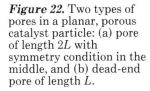
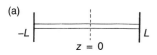
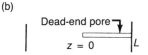

The defining steady-state continuity-of-species equation contains only two terms,

$$-D_{is}\frac{d^2 c_{is}}{dz^2} = R_{is} \tag{66}$$

with two boundary conditions:

$$\text{B.C. 1:} \quad -D_{is}\frac{dc_{is}}{dz} = 0 \quad \text{at } z = 0 \tag{67}$$

$$\text{B.C. 2:} \quad c_{is} = c_{is}^0 \quad \text{at } z = L \tag{68}$$

Observe that boundary condition 1 is applicable for either of two situations: a pore of length $2L$ and a symmetry situation in the middle of the pore, and a pore of length L that has an impermeable end at $z = 0$. The first condition is the more reasonable one for catalyst pellets.

The solution to Equation 66 for a first-order rate law, Equation 59, is

$$c_{is} = c_{is}^0 \frac{\cosh z\sqrt{k_{is}/D_{is}}}{\cosh L\sqrt{k_{is}/D_{is}}} \tag{69}$$

and the corresponding concentration profile within the pore is shown in Figure 23. In deriving Equation 69, we have taken several liberties; namely, we have converted a surface rate constant (for the catalyst deposited on the pore walls) into a volumetric rate constant. The basis for this conversion will be discussed in the section, "Multiphase/Multistate/Coupled Chemical Systems."

Figure 23. Concentration profiles, for different values of the Thiele modulus, in a steady-state diffusion-controlled reaction system. See Equation 69.

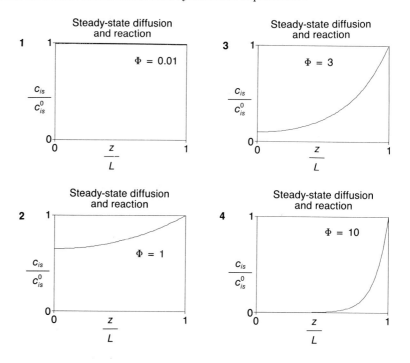

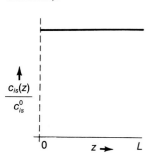

Figure 24. Concentration profile when the diffusion rate through the pore is much greater than the reaction rate (small Thiele modulus).

What happens when the pore length, L, is small; the first-order rate constant, k_{is}, is small; and the diffusion coefficient, D_{is}, is large? The concentration profile becomes flat throughout the pore (Figure 24). Why? Because the rate of diffusion is sufficiently large to supply fresh reactant i even to the center (or end) of the pore. A flat concentration profile corresponds to the maximum rate of reaction within the pore.

We are not finished. An extremely important theoretical result occurs when we compute the ratio of the actual total rate of reaction in the pore to the maximum total reaction possible. This calculation involves integration of the hyperbolic concentration profile as follows:

$$\frac{\text{Actual total rate}}{\text{Maximum total rate}} = \frac{\int_0^L k_{is} c_{is} dz}{\int_0^L k_{is} c_{is}^0 dz}$$

$$= \frac{\int_0^L \cosh z \sqrt{k_{is}/D_{is}} \, dz}{L \cosh L \sqrt{k_{is}/D_{is}}}$$

$$= \frac{\sinh L \sqrt{k_{is}/D_{is}}}{L \sqrt{k_{is}/D_{is}} \cosh L \sqrt{k_{is}/D_{is}}}$$

$$= \frac{\tanh L \sqrt{k_{is}/D_{is}}}{L \sqrt{k_{is}/D_{is}}}$$

$$= \frac{\tanh \Phi}{\Phi} \tag{70}$$

The quantity, Φ, is the famous dimensionless group, $L\sqrt{k_{is}/D_{is}}$, that in the United States is known as the Thiele modulus (22). All chemical engineers know (or should know) the significance of this group, and so should most chemists!

We asked what happens when the pore length, L, is small; the first-order rate constant, k_{is}, is small; and the diffusion coefficient, D_{is}, is large. The terms "small" and "large" are not very quantitative. The Thiele modulus permits us to ask this group of three questions in a more precise manner: What happens when the Thiele modulus is small? Figure 25 provides a plot of $\tanh \Phi/\Phi$ versus the value of Φ.

Figure 25. Plot of the effectiveness factor (ratio of the hyperbolic tangent of the Thiele modulus to the Thiele modulus) as a function of Thiele modulus. Significant diffusion control begins to occur when the Thiele modulus is greater than 1. See Equation 70.

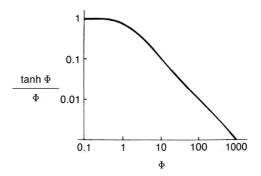

Observe that at small values of Φ the ratio of rates in Equation 70, called the *effectiveness factor*, is unity. This means that when the rate of reaction within the pore is much smaller than the rate of diffusion through the pore, the concentration profile is flat and the maximum total reaction rate occurs. On the other hand, as Φ becomes larger, the ratio of reaction to diffusion rates becomes sufficiently large that most of the reaction occurs very near the surface ($z = L$) of the pore. This means that *much of the catalyst surface in the interior pore is not used; it is wasted*! Imagine having a large industrial catalytic reactor (a pipe packed with porous catalyst pellets) in which the inventory of noble metal, for example, platinum, is worth $1 million and only 10% of the platinum (that near the surface of each pellet) is actually used to perform the reaction. We conclude that steady-state diffusion and reaction in catalyst pellets has substantial economic consequences if not performed properly.

The reader should consult Reference 4 for a variety of interdisciplinary examples of steady-state diffusion and reaction.

Steady-State Diffusion and Convection
The defining equation for steady-state diffusion and convection is

$$-D_{is}\frac{d^2 c_{is}}{dz^2} + v_s \frac{dc_{is}}{dz} = 0 \qquad (71)$$

This equation becomes interesting primarily when there is more than a single moving phase in a chemical system. The mathematical treatment of systems of this kind will be deferred to a problem in the section "Multistate/Multiphase/Coupled Chemical Systems."

Non-Steady-State Diffusion, Convection, and Reaction
The defining equation for non-steady-state diffusion, convection, and reaction is the full continuity-of-species equation itself:

$$\underbrace{\frac{\partial c_{is}}{\partial t}}_{\text{accumulation}} \underbrace{- D_{is}\frac{\partial^2 c_{is}}{\partial z^2}}_{\text{diffusion}} + \underbrace{v_s \frac{\partial c_{is}}{\partial z}}_{\text{convection}} = \underbrace{R_{is}}_{\text{reaction}} \qquad (1)$$

The transformation given by Equation 47 is useful here, as well. The result is

$$\left(\frac{\partial c_{is}}{\partial t}\right)_{u_s} - D_{is}\left(\frac{\partial^2 c_{is}}{\partial u_s^2}\right)_t = R_{is} \tag{72}$$

which, as one example of a chemical system (Figure 26), can be solved for initial and boundary conditions given by Equation 56. The solution for a first-order rate law

Figure 26. Non-steady-state diffusion, convection, and reaction in a tubular flow system.

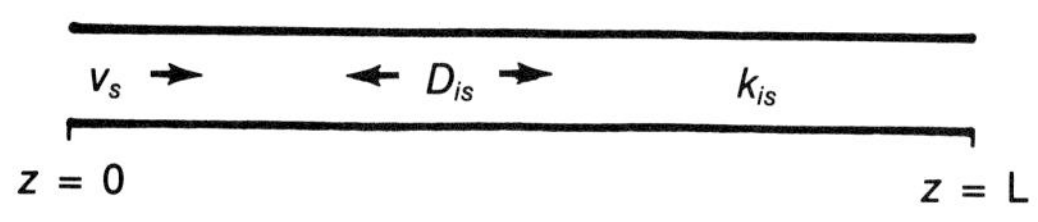

(Equation 59) can be obtained readily through the use of the following transformation,

$$c_{is}(u_s,t) = e^{-k_{is}t} f_{is}(u_s,t) \tag{73}$$

where $f_{is}(u_s,t)$ is the solution to the equation

$$\left(\frac{\partial f_{is}}{\partial t}\right)_{u_s} - D_{is}\left(\frac{\partial^2 f_{is}}{\partial u_s^2}\right)_t = 0 \tag{74}$$

The final result is

$$c_{is}(z,t) = \frac{n_{is}^0 e^{-k_i t}}{A_{\perp s}\sqrt{\pi D_{is} t}} \exp\left[-\frac{(z - v_s t)^2}{4 D_{is} t}\right] \tag{75}$$

Compare the profiles in Figure 27 with those in Figure 17. In this case, not only does the peak broaden, it decreases in magnitude and ultimately disappears as the reactant i is converted to product. Whether or not there is any i left at $z = L$ depends upon the relative rates of reaction and convection in the tubular non-steady-state reactor.

Figure 27. Concentration profiles, as a function of time, for a diffusing and reacting solute pulse as it moves through a tubular flow system. See Equation 75.

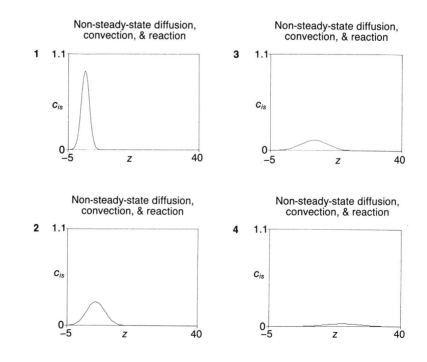

Non-Steady-State Convection with Lateral Fields: The Principle of Field-Flow Fractionation

Field-flow fractionation (23) is an elegant technique that is based upon the existence of lateral velocity (Figure 28) and concentration (Figure 29) profiles in a homogeneous chemical system that is flowing between, for example, a pair of flat plates that are very close to each other and have some type of field between them. This field—thermal, electric, magnetic, and so forth—creates the concentration profile shown in Figure 29. In the simplest continuity-of-species equation, diffusion is assumed to be negligible (not always a realistic assumption), and the equation becomes

$$A_{\perp s}\frac{\partial c_{is}}{\partial t} - \frac{\partial}{\partial z}\iint_{A_{\perp s}} v'_s(A_{\perp s})c_{is}(A_{\perp s})dA_{\perp s} = 0 \tag{76}$$

Different solutes i create profiles of different slopes within the field, and thus feel the parabolic velocity profile (Figure 28) differently. In this way, chromatographic-type separations are achieved in a homogeneous phase. For a rectangular cross-sectional area of width W, thickness L, and distance variable x_s, $A_{\perp s} = WL$, $dA_{\perp s} = Wdx_s$, and $v'_s c_{is} = v'_s(x_s)c_{is}(x_s)$.

Figure 28. Parabolic velocity profile for a fluid that flows between two parallel plates.

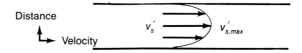

Figure 29. Lateral concentration profile produced by a field (thermal, electric, gravitational, or magnetic) in a field-flow fractionation apparatus.

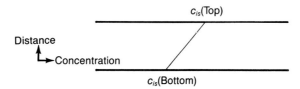

The continuity-of-species equation for a linear multistate chemical system

Heterogeneous Systems: The Concept of a State

Most chemical systems in practical situations are heterogeneous systems. Even the classic homogeneous batch reactor in organic chemistry—the three-necked round-bottom flask—is a heterogeneous system in principle. In addition to the stirred liquid, there are several "phases," namely, the glass walls, the surface of the stirrer, and the gas above the liquid. Only under the assumption of no equilibration or reaction on the walls and in the gas does the assumption of homogeneity become valid for the system.

Some sort of bookkeeping is required to keep track of the different solutes i in the different phases of a heterogeneous system. We have anticipated this need by already defining a phase subscript, s, and using it consistently in our conservation-of-species equations, even when only a single phase was present. Thus, the concentration of solute i in phase s was written as c_{is}. The corresponding definitions of R_{is}, D_{is}, and v_s were listed following Equation 1.

We call the combination of solute i in phase s a *state* (17 years ago, we called such a combination a *partition state*) (9, 24, 25). Such nomenclature simplifies the discussion of simultaneous reaction, diffusion, and convection in heterogeneous systems in which rapid equilibration of solutes i among the phases s occurs. The reader is referred to References 24 and 25 for a discussion of the thermodynamic significance of the concept of a state. The customary statement of phase equilibrium (*24*), "If two phases are in equilibrium, all components capable of passing from one to the other must have the same chemical potential in the two phases," can be broadened, through the use of the concept of a state, to incorporate both physical (i.e., phase) and chemical equilibria: "If two states are in equilibrium, a component capable of passing from one to the other must have the same chemical potential in the two states." This is an important generalization, because it permits us to write the continuity-of-species equation for a class of heterogeneous chemical systems in a manner that is independent of the nature of the equilibria that occur.

It should be emphasized a final time that the concept of a state has both bookkeeping and thermodynamic significance. In this chapter, it is the bookkeeping function that is most useful. If a solute i can be identified by physical or chemical means, and if it is present in/at different phases or interfaces s, then each combination *is* represents a different state for the solute. In terms of the continuity-of-species equation, the dynamic behavior of the solute can be characterized in terms of a diffusion coefficient, velocity, and rate constant for each state.

The Continuity-of-Species Equation for a State

The continuity-of-species equation for solute i in a single state *is* has been given previously as the three-dimensional equation (Equation 2):

$$\underbrace{\frac{\partial c_{is}}{\partial t}}_{\text{accumulation}} - \underbrace{D_{is}\nabla^2 c_{is}}_{\text{diffusion}} + \underbrace{v_s \nabla c_{is}}_{\text{convection}} = \underbrace{R_{is}}_{\text{reaction}} \qquad (2)$$

Three-dimensional partial differential equations are not easy to solve, and they tend to mask the principles that underlie physical situations. It would be desirable to simplify Equation 2 to a single dimension. As will be discussed later, such an objective can be readily achieved by subdividing the three-dimensional coordinate system into two classes of coordinate directions: (1) two lateral coordinate directions, in which the combination of an impermeable lateral boundary and rapid equilibrium of solute i among its states eliminates lateral concentration gradients, and (2) a single axial coordinate direction in which dynamic processes, such as transport, create one-dimensional spatial concentration gradients. Such a system is an example of *two-dimensional thermodynamics* superimposed upon *one-dimensional dynamics*. The thermodynamics occurs in the two lateral coordinate directions, and the dynamics occurs in the axial coordinate direction. The mathematics of such a system is strictly one-dimensional! Therefore, the simple theoretical expedient of separating chemical processes among coordinate directions has significant pedagogical value.

Equation 2 can be rewritten to acknowledge the distinction between the lateral and axial coordinate directions. The resulting equation has one more term than Equation 2:

$$\underbrace{\frac{\partial c_{is}}{\partial t}}_{\text{accumulation}} - \underbrace{D_{is}\frac{\partial^2 c_{is}}{\partial z^2}}_{\substack{\text{axial}\\\text{diffusion}}} + \underbrace{v_s \frac{\partial c_{is}}{\partial z}}_{\substack{\text{axial}\\\text{convection}}} + \underbrace{\nabla \cdot \mathbf{N}_{\perp is}}_{\substack{\text{lateral}\\\text{transport}}} = \underbrace{R_{is}}_{\text{reaction}} \qquad (77)$$

Equation 77 is written for each state *is* in the chemical system. For a given solute i and three states (for example, the solute in the gas phase, liquid phase, and interface), three such equations would be required to characterize the system.

The requirement to solve several simultaneous three-dimensional partial differential equations is a formidable undertaking. Can Equation 77 be simplified? No. Can the set of simultaneous equations be reduced in number? The answer is yes, and the gimmick is *two-dimensional thermodynamics*. How can we convert the three-dimensional Equation 77 into a one-dimensional equation? We simply find a method to eliminate the lateral transport term, $\nabla \cdot \vec{N}_{\perp is}$, in all equations of a set of Equations 77. By doing so, we reduce this set of equations to a form similar to Equation 1, a significant result. Further details are in the next subsection.

The Continuity-of-Species Equation for a Linear Multistate Chemical System

The task before us is to simplify the three-dimensional partial differential Equation 77 that characterizes the continuity-of-species behavior of solute i in state is in a chemical system. Unfortunately, we cannot simplify a single Equation 77, for example, by eliminating the lateral transport term. Instead, we must simplify the entire set of three-dimensional partial differential Equations 77 that characterize the behavior of a single solute i in all of its states $is = i1, i2, i3, \ldots$ within the chemical system.

Space does not permit a thorough discussion of how we do this. For full details of the derivation, see References 7 and 9. The key assumptions underlying the derivation can be stated as follows:

1. Two or more states exist for solute i. It will be shown later that the existence of two or more states need not imply that the chemical system is heterogeneous.
2. The chemical system consists of a single axial coordinate (typically the z-axis in Cartesian or cylindrical coordinate systems), along which transport processes can occur and spatial concentration gradients can exist, and two lateral coordinates, along which rapid equilibrium processes occur and no spatial concentration gradients exist.
3. The net rate of mass transfer of species i in the lateral coordinate directions is zero, either because of an impermeable lateral boundary or because of a symmetry condition.
4. Solute i exists only as a monomer in any of its states. (**Note:** We make this assumption to eliminate nonlinear partial differential equations. Such equations usually do not enhance the pedagogical value of the theory.)
5. Linear partition coefficients characterize the concentration of solute i in any of its states. (*Note:* We make this assumption to eliminate nonlinear partial differential equations.)
6. The rate of reaction is first order in solute i in any state where a reaction occurs. (*Note:* We make this assumption to eliminate nonlinear partial differential equations.)
7. The density of each phase s is constant.
8. The diffusion coefficients, velocities, and rate constants are not functions of the axial coordinate or time.

Assumptions 4, 5, 6, and 7 can all be relaxed, but at the cost of a substantially more complex mathematical model. For a junior-level physical chemistry course, the increase in generality does not justify the increase in complexity, which should be avoided as much as possible.

The derivation proceeds as follows: The continuity-of-species equation (Equation 77) is written for each state is. The s equations are summed and then integrated over the cross-sectional area of the system (9). Green's theorem is employed to eliminate the lateral mass transport term for solute i (note that this theorem is applied to the summed set of equations, not to a single Equation 77 (9). The result is a one-dimensional partial differential equation,

$$\sum_{s=1}^{n} A_{\perp s} \left(\frac{\partial c_{is}}{\partial t} - D_{is} \frac{\partial^2 c_{is}}{\partial z^2} + v_s \frac{\partial c_{is}}{\partial z} - R_{is} \right) = 0 \tag{78}$$

that contains a summation over all states is. The quantity $A_\perp$ is the total cross-sectional area of the chemical system, and the quantity $A_{\perp s}$ is the total cross-sectional area of a given phase s (Figure 30).

The next step is the assumption of a linear relationship between the concentration of solute i in state is and its concentration in an arbitrarily chosen reference state $isref$. The constant that characterizes this relationship is known in the chemical literature as a *partition coefficient*,

$$\kappa_{is} = \frac{c_{is}}{c_{isref}} \tag{79}$$

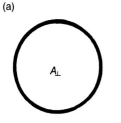

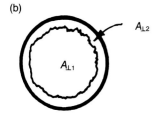

Figure 30. Cross-sectional areas in (a) a single-phase, and (b) a two-phase chemical system. It is assumed that such areas remain constant throughout the entire length of the one-dimensional system.

This equation is the essence of our assumption of *two-dimensional thermodynamics*. It means that at any given axial coordinate value z, and at any time t, all of the states *is* are in equilibrium with each other. A change in concentration c_{is} in one of the states instantaneously influences the concentrations in all other states according to the partition coefficient relationship (Equation 79).

The result of applying Equations 59 and 79 to equation 78 is

$$\frac{\partial c_{isref}}{\partial t} - \frac{\partial^2 c_{isref}}{\partial z^2} \frac{\sum_{s=1}^{n} A_{\perp s} \kappa_{is} D_{is}}{\sum_{s=1}^{n} A_{\perp s} \kappa_{is}} + \frac{\partial c_{isref}}{\partial z} \frac{\sum_{s=1}^{n} A_{\perp s} \kappa_{is} v_{is}}{\sum_{s=1}^{n} A_{\perp s} \kappa_{is}} + c_{isref} \frac{\sum_{s=1}^{n} A_{\perp s} \kappa_{is} k_{is}}{\sum_{s=1}^{n} A_{\perp s} \kappa_{is}} = 0 \quad (80)$$

which can be simplified to

$$\underbrace{\frac{\partial c_{isref}}{\partial t}}_{\text{accumulation}} - \underbrace{D_{ieff} \frac{\partial^2 c_{isref}}{\partial z^2}}_{\substack{\text{multistate}\\\text{diffusion}}} + \underbrace{v_{ieff} \frac{\partial c_{isref}}{\partial z}}_{\substack{\text{multistate}\\\text{convection}}} + \underbrace{k_{ieff} c_{isref}}_{\substack{\text{multistate}\\\text{reaction}}} = 0 \quad (81)$$

The physical significance of the constant "effective" parameters D_{ieff}, v_{ieff}, and k_{ieff} will be discussed in the following sections. Since any state can be selected as *isref*, the "ref" can be dropped from Equation 81.

A chemical system that satisfies the above assumptions, as well as others associated with the derivation, is called a *linear multistate chemical system* (9). Ideal thermodynamic systems exhibit no spatial concentration gradients in any of the *three coordinate directions*. The linear multistate chemical system approach relaxes such a restriction to only *two coordinate directions* (the lateral coordinates), with the third coordinate direction having dynamic processes that are not at thermodynamic equilibrium. Stated in another way, Equation 81 is a theory of the dynamics of two-dimensional thermodynamic systems.

Effective Diffusion Coefficient, Velocity, and Rate Constant

The effective diffusion coefficient, velocity, and rate constant are defined by the following three equations, all of which have the same basic form and represent weighted averages of the diffusion coefficient, velocity, and rate constant of solute i in all of its equilibrating states *is*:

$$D_{ieff} = \sum_{s=1}^{n} Y_{is} D_{is} \quad (82)$$

$$v_{ieff} = \sum_{s=1}^{n} Y_{is} v_s \quad (83)$$

$$k_{ieff} = \sum_{s=1}^{n} Y_{is} k_{is} \quad (84)$$

In these equations,

D_{is} = diffusion coefficient of solute i in state *is* (length²/time)
v_s = mass-average velocity of solute i in state *is* (length/time)
k_{is} = first-order rate constant of solute i in state *is* (1/time)
Y_{is} = fraction of solute i in state *is* (dimensionless)
c_{is} = concentration of solute i in state *is* (mol/volume)
t = time
z = axial coordinate direction (length)

Segregation Fraction

The diffusion coefficient, velocity, and rate constant have already been explained in previous sections of this chapter. The only new quantity in Equations 82–84 is the *segregation fraction* of solute i in state *is*, Y_{is}, which is a dimensionless quantity (25).

In a system at thermodynamic equilibrium, Y_{is} is defined mathematically as

$$Y_{is} = \frac{n_{is}}{n_i^0} \quad (85)$$

where n_{is} is the number of moles of solute i in state is, and n_i^0 is the total number of moles of i in the system. What happens in a two-dimensional thermodynamic system? For a linear system with linear distribution coefficients, the segregation fraction is determined according to the equation

$$Y_{is} = \frac{K_{is}}{\sum_{s=1}^{n} K_{is}} \tag{86}$$

where K_{is} is a distribution coefficient, also a dimensionless quantity. Equation 86 also applies to Equation 85 in systems with linear distribution coefficients and provides an alternative method for computing Y_{is}.

Distribution Coefficient

In a system at thermodynamic equilibrium, the *distribution coefficient* is defined mathematically as the ratio of the number of moles of solute i in state is to the number of moles of solute i in a reference state, which we have designated as $isref$:

$$K_{is} = \frac{n_{is}}{n_{isref}} \tag{87}$$

Note that K_{is} is a dimensionless quantity. The selection of the reference state is arbitrary and generally follows customs established in the literature.

What happens in a two-dimensional thermodynamic system? The distribution coefficient is written as the product of a partition coefficient, κ_{is}, and either a volume fraction, ϵ_s, or a surface-to-volume ratio, σ_s:

$$K_{is} = \kappa_{is} \frac{\epsilon_s}{\epsilon_{sref}} \tag{88}$$

$$K_{is} = \kappa'_{is} \frac{\sigma_s}{\epsilon_{sref}} \tag{89}$$

These additional physical parameters will be discussed next.

Partition Coefficient

Once the concept of a state has been defined, the distinction between the concepts of *partition* coefficient and distribution coefficient can be given without difficulty:

partition coefficient The ratio of the concentrations of a solute i in two of its states.

distribution coefficient The ratio of the amounts of a solute i in two of its states.

Whereas the distribution coefficient is defined as Equation 87, the partition coefficient is defined as

$$\kappa_{is} = \frac{c_{is}}{c_{isref}} \text{ or } \kappa'_{is} = \frac{c'_{is}}{c_{isref}} \tag{79}$$

A prime is used for both the concentration c'_{is} and the partition coefficient κ'_{is} if the state is is a surface state. In such a case, the surface concentration has units of mol/area, and the partition coefficient has units of vol/area.

Why is it necessary to define a reference state? Consider the following example. In a heterogeneous system of five states ($is = i1, i2, i3, i4$, and $i5$), there are four independent partition coefficients for a solute i that equilibrates between the states. Being a ratio of concentrations, the partition coefficient must have the concentration of one of these five states in the denominator. In order to simplify computations, this concentration should be in the denominator in all four partition states. We call the state whose concentration is in the denominator the *reference state* and use the notation *isref* to characterize it. Thus, for $isref = i1$, the four independent partition coefficients in the five-state system are

$$\kappa_{i2} = \frac{c_{i2}}{c_{i1}} \tag{90}$$

$$\kappa_{i3} = \frac{c_{i3}}{c_{i1}} \tag{91}$$

$$\kappa_{i4} = \frac{c_{i4}}{c_{i1}} \tag{92}$$

$$\kappa_{i5} = \frac{c_{i5}}{c_{i1}} \tag{93}$$

In linear systems, the partition coefficient is constant for any value of c_{is} or c_{isref}. Concentration can have units of either mol/vol (for a volumetric state) or mol/area (for a surface state). To repeat, a surface concentration has a prime associated with it, as in c'_{is}. In order to calculate the amount of solute i in a given state is, n_{is}, one must multiply the concentration by either a volume v_s or an area S_s:

$$n_{is} = c_{is} V_s \tag{94}$$

$$n_{is} = c'_{is} S_s \tag{95}$$

Unlike the distribution coefficient, which is dimensionless, the partition coefficient has dimensions of either volume/volume or else volume/area. Different phases are like apples and oranges; their volumes must be retained when units for symbols are given. As in any good bookkeeping system, retention of the units of the phases under consideration provides an important reminder that the total amounts of the phases will be required in the course of the calculation.

In this chapter, the symbol for a partition coefficient is the Greek letter kappa, κ, plus a suitable state subscript. A better symbol than kappa is desirable to represent partition coefficients because of the potential confusion among the symbols for partition coefficient, κ, distribution coefficient, K, and rate constant, k. Kappa is retained here because of its prior use in the literature (9, 24, 25).

Volume Fraction and Surface-to-Volume Ratio

Chemistry textbooks do not treat heterogeneous systems as commonly as do chemical engineering textbooks. As a result, chemistry students fail to gain an appreciation for two very important quantities in heterogeneous systems, namely, the volume fraction and the surface-to-volume ratio. Definitions are in order:

volume fraction   The ratio of the volume of phase s to the total volume of the chemical system.

surface-to-volume ratio   The ratio of the surface area between two phases to the total volume of the chemical system.

Neither quantity is dimensionless because volume units associated with different phases do not cancel.

The manner in which the volume fraction appears in the derivation of the continuity-of-species equation for a linear multistate chemical system is as a ratio of cross-sectional areas of the chemical system, which are assumed to be uniform as a function of the axial spatial variable z,

$$\epsilon_s = \frac{A_{\perp s}}{A_{\perp}} \tag{96}$$

Two other, more standard definitions for the volume fraction and the surface-to-volume ratio are (9):

$$\epsilon_s = \frac{V_s}{V} \tag{97}$$

$$\sigma_s = \frac{S_s}{V} \tag{98}$$

Note that a volume is the product of a cross-sectional area times a length (for example, the total length L of the system):

$$V = A_{\perp} L \tag{99}$$

$$V_s = A_{\perp s} L \tag{100}$$

A surface-to-volume ratio is an important quantity because it significantly influences the total rate of mass transfer (mol/s) between two phases, or between a phase and an interface (as would occur with a heterogeneous catalyst). The higher the surface-to-volume ratio, the greater the mass transfer rate. Two examples where high ratios are important are in the human lungs and in porous heterogeneous catalysts.

Physical and Chemical Equilibria
Several definitions are in order (24):

reaction   Taken in its broadest sense, any change in the internal constitution of a system. Such a change may be in the physical state of the components or it may be the result of chemical transformations among the molecules in the system (26).

physical reaction   A reaction in which the change is in the physical state of the components.

chemical reaction   A reaction in which the change is a chemical transformation among the molecules.

When applied to chemical systems, the term equilibrium has in the past been used in two ways: to denote the condition of equilibrium, in which the chemical affinity of each reaction is equal to zero (12, page 41), and to collectively denote physical or chemical reactions whose affinities are equal to zero (for example, when the terms phase equilibria or chemical equilibria are used in textbooks and articles). We shall follow the second of these meanings and use the following definitions (24):

chemical equilibrium   A chemical reaction in which the chemical affinity is equal to zero.

physical equilibrium   A physical reaction in which the chemical affinity is equal to zero.

Thus we use the term physical equilibria instead of phase equilibria, a term frequently found in the literature.

Summary

We have defined the concept of a state to facilitate discussions of "heterogeneous" chemical systems in which two-dimensional physical and chemical equilibria are superimposed on one-dimensional dynamic processes such as diffusion, convection, and reaction. The result is a simple and powerful equation, Equation 81, that can be used to teach the basic principles of chromatography, continuous separation processes, facilitated diffusion, heterogeneous catalytic reactions, and other processes that rely upon the continuity-of-species equation. As we will show in the following section, we have reduced the theoretical treatment of multistate chemical systems to (1) solution of the homogeneous one-dimensional continuity-of-species equation, and (2) bookkeeping associated with the many states in which a solute i can participate. We have made no fundamental distinction between physical and chemical equilibria in defining the concept of a state.

Multiphase/multistate/coupled chemical systems

Non-Steady-State Multistate Diffusion and Convection: The Principle of Capillary Chromatography

Nowhere in the chemical sciences but in the field of chromatography is the value of the linear multistate chemical system approach clearer (9). The defining equation is an example of non-steady-state multistate diffusion and convection,

$$\frac{\partial c_{is}}{\partial t} - D_{ieff}\frac{\partial^2 c_{is}}{\partial z^2} + v_{ieff}\frac{\partial c_{is}}{\partial z} = 0 \tag{101}$$

which is identical in form to Equation 54 for the broadening of a peak in a flowing system. The solution, Equation 102, is identical in form to Equation 57 for the injection of an instantaneous pulse at $z = 0$ and $t = 0$ (see Equation 56 and Figure 31):

$$c_{is}(z,t) = \frac{n_{is}^0}{A_{\perp s}\sqrt{\pi D_{ieff} t}} \exp\left[-\frac{(z - v_{ieff} t)^2}{4 D_{ieff} t}\right] \tag{102}$$

The time required to move component i from $z = 0$ to $z = L$ is called the *retention time*, t_{Ri},

Figure 31. Concentration profiles, as a function of time, for a capillary chromatographic separation system. See Equation 102.

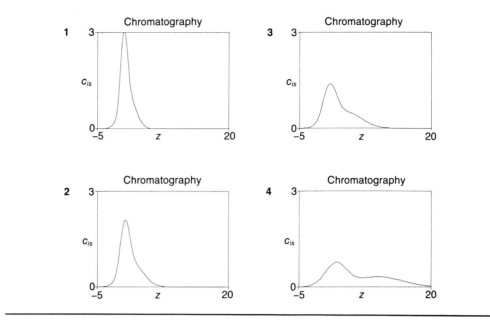

$$L = \int_0^{t_{Ri}} v_{ieff}\, dt \tag{103}$$

(Recall that it was called the *residence time* in Equation 53.) When we substitute one of the forms for the effective velocity, v_{ieff},

$$v_{ieff} = \frac{\sum_{s=1}^{n} K_{is} v_s}{\sum_{s=1}^{n} K_{is}} \tag{104}$$

and assume that v_{ieff} is constant as a function of distance and time, we obtain

$$L = t_{Ri} \frac{\sum_{s=1}^{n} K_{is} v_s}{\sum_{s=1}^{n} K_{is}} \tag{105}$$

which is one of the fundamental relationships in the field of chromatography.

As an example of the application of Equation 105, consider a two-state system in which Phase 1 is the gas phase and Phase 2 is a liquid phase deposited as a thin film on the inner surface of a long tubular column (capillary chromatography). The liquid, of course, is stationary and has no velocity. Thus, the retention time is simply

$$t_{Ri} = \frac{L(1 + K_{i2})}{v_1} \tag{106}$$

The reference state is chosen to be the solute i in the gas phase, or $i1$. Note that the distribution coefficient for the reference state is unity by definition:

$$K_{i1} = \frac{n_{i1}}{n_{i1}} \equiv 1 \tag{107}$$

References 9 and 27 explored the applications of Equation 101 to a wide variety of chromatography systems. Perhaps the most interesting conclusion that resulted from this work was the fact that *chromatography need not be a two-phase technique*. *Carrier electrochromatography*, in which chromatographic action occurs by mobile, soluble complexing agents in a homogeneous solution contained within an electrochemical cell, is entirely consistent with the fundamental principles of the field

(9) and may ultimately become a technique that is used commercially in the Spacelab. Other unusual chromatographic techniques include carrier magnetochromatography, inverse carrier electrochromatography, solid-phase carrier electrochromatography (9), dust chromatography, aerosol chromatography, and so forth (27).

Non-Steady-State Multistate Reaction: The Principle of Homogeneous Catalysis
One equation for a homogeneous catalytic reaction in a batch reactor is

$$\frac{dc_{is}}{dt} + k_{ieff} c_{is} = 0 \tag{108}$$

where it is assumed that the reactions are all pseudo-first-order reactions. For the same initial condition as in Equation 60, the solution is

$$c_{is} = c_{is}^0 e^{-k_{ieff} t} \tag{109}$$

which, of course, is identical in form to Equation 61.

The interesting parameter in this chemical system is the effective first-order rate constant, which can be written in terms of distribution coefficients:

$$k_{ieff} = \frac{\sum_{s=1}^{n} K_{is} k_{is}}{\sum_{s=1}^{n} K_{is}} \tag{110}$$

As an example, consider a two-state system in which $i1$ is the reference state (the free solute i in solution), and $i2$ is a soluble, reactive complex in which i is complexed with a catalyst such as a homogeneous metal complex. State $i1$ is unreactive, so k_{i1} is zero. The effective rate constant is, therefore,

$$k_{ieff} = \frac{K_{i2} k_{i2}}{1 + K_{i2}} \tag{111}$$

where it should be kept in mind that the extent of complexing is assumed to be such that the reaction rate remains first order. The significance of the value of the distribution coefficient in dictating the magnitude of k_{ieff} is clear.

Steady-State Multistate Convection and Reaction: The Principle of the Fixed-Bed Catalytic Reactor
The defining equation for a steady-state heterogeneous fixed-bed catalytic reactor, that is, a cylindrical tube packed with a porous catalytic solid, is

$$v_{ieff} \frac{dc_{is}}{dz} + k_{ieff} c_{is} = 0 \tag{112}$$

For the boundary condition given by Equation 63, the solution is

$$c_{is} = c_{is}^0 e^{-k_{ieff} z / v_{ieff}} \tag{113}$$

The most interesting quantity that characterizes such a system is the dimensionless quantity $k_{ieff} L / v_{ieff}$, which can be written in terms of distribution coefficients as

$$\frac{k_{ieff} L}{v_{ieff}} = L \frac{\sum_{s=1}^{n} K_{is} k_{is}}{\sum_{s=1}^{n} K_{is} v_s} \tag{114}$$

Consider a three-state system that consists of a gas reference state $i1$, an inert porous solid $i2$, and a catalytic surface $i3$ within the porous solid. Reaction occurs only on the catalytic surface, which has a surface-to-volume ratio, σ_3, relative to the total volume of the reactor. Equation 114 can be reduced to

$$\frac{k_{ieff} L}{v_{ieff}} = L \frac{k_{i1} + K_{i2} k_{i2} + K_{i3} k_{i3}}{v_1 + K_{i2} v_2 + K_{i3} v_3} \tag{115}$$

where no decisions so far have been made concerning which terms should be retained.

Now the fun begins. No reaction occurs in either state $i2$ or $i3$, so two of the three terms in the numerator of Equation 115 disappear. Also, only the gas is moving through the fixed-bed reactor; states $i2$ and $i3$ are stationary. Two more terms disappear from the denominator. The final result is

$$\frac{k_{ieff}L}{v_{ieff}} = \frac{K_{i3}k_{i3}L}{v_1} \tag{116}$$

This process is simply an exercise in physical and chemical bookkeeping applied to the physical and dynamic characteristics of the system.

We are not yet finished. With the aid of the definition for the surface distribution coefficient,

$$K_{i3} = \kappa'_{i3}\frac{\sigma_3}{\epsilon_1} \tag{117}$$

where σ_3 is a surface-to-volume ratio, Equation 116 becomes

$$\frac{k_{ieff}L}{v_{ieff}} = \kappa'_{i3}k_{i3} \cdot \frac{\sigma_3 L}{\epsilon_1 v_1} = \kappa'_{i3}k_{i3} \cdot \frac{S_3 L}{V_1 v_1} = \kappa'_{i3}k_{i3} \cdot \frac{S_3}{F_1} \tag{118}$$

which is a fundamental result in chemical engineering. Note that $F_1 = v_1 V_1/L$. Equation 118, when combined with the solution, Equation 113, states that if the following plot is made,

$$\ln\frac{c_{i1}}{c^0_{i1}} \text{ vs. } \frac{1}{F_1} \tag{119}$$

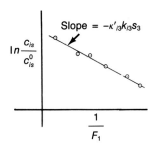

Figure 32. Data analysis for a fixed-bed catalytic reactor in which the reaction is first order in reactant. Observe that the slope provides a product of physical quantities rather than the rate constant directly. See Equations 113 and 118.

the product of the rate constant, surface partition coefficient, and total catalytic surface area in the reactor can be measured (Figure 32). A knowledge of the total surface area permits computation of the product, $\kappa'_{i3}k_{i3}$, which is about as far as we can go in kinetic measurements.

The quantity, $\kappa'_{i3}k_{i3}$, is what we measure in a fixed-bed catalytic reactor. The units of this quantity are interesting:

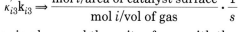

If we simply cancel the units of area with the units of volume, we conclude that the experimentally measured result, $\kappa'_{i3}k_{i3}$, has units of cm · s. But how can a first-order rate constant have units that involve length? The fact that a surface partition coefficient, which is a measure of the solubility of the solute on the catalyst surface, is also present in the experimentally measured result provides a satisfactory answer, one that is frequently missed by both undergraduate and graduate students.

For further study: Problems and challenges associated with multiphase/multistate/coupled chemical systems

We have provided only three examples of multiphase/multistate/coupled chemical systems. Many more exist. Much of the enjoyment associated with the application of the "linear multistate chemical system" approach resides in the discovery of new systems where the principles apply in a simple and direct manner. To assist you and your student colleagues in such discovery, we provide the following problems.

Steady-State Multistate Diffusion: Transport through a Plane Membrane with an Immobile Complexing Agent
Consider a solute i that permeates through a planar polymeric membrane. Within the membrane, two states for the solute exist: a "mobile" state that represents the freely diffusing solute, and a "stationary" state that represents the solute complexed to a reversible complexing agent that is located on essentially immobile polymer chains.

What is the effective diffusion coefficient for the solute, and how does it depend upon the partition coefficient and the concentration of the immobile complex within the polymer? Choose the solute i in the liquid external to the polymer as the reference state.

Steady-State Multistate Diffusion: Carrier Transport through a Plane Membrane
Consider a solute i that permeates through a planar polymeric membrane. Within the membrane, two states for the solute exist: a state that represents the freely diffusing

solute, and a mobile complexed state that represents the solute reversibly complexed to a *mobile carrier molecule*, which itself is transported at a steady rate through the plane membrane. In other words, (1) the solute complexes with the carrier on one side of the membrane, (2) the carrier-solute reversible complex diffuses through the membrane, and (3) the complex releases the solute on the other side of the membrane.

What is the effective diffusion coefficient for the solute, and how does it depend upon the partition coefficient and the concentration of the carrier complex within the polymer? Choose the solute i in the liquid external to the polymer as the reference state.

Steady-State Multistate Diffusion and Convection: The Principle of Countercurrent Separations

Have your students ever wondered why chromatography seemingly provides "much better" separations than the older steady-state separation techniques such as distillation and extraction? Just how good can ideal countercurrent techniques be? One way to tackle such questions is through the following problem.

Consider a continuous *countercurrent extraction system* consisting of two separate columns joined at the middle, where the feed of solutes $i = 1, 2, 3, \ldots$ enters. By analogy to distillation, there is a stripping section on one side of the feed and a rectification section on the other side. Assume that analogs to the distillation-column condenser and reboiler exist at each end of the extraction system to provide reflux, which is the fundamental "gimmick" of countercurrent separation systems. Stated in another way, the gimmick of reflux involves two steady-state countercurrent fluid streams that permit the feed solutes to continuously equilibrate between the countercurrent streams as a function of distance.

Consider a small amount of solute i continuously injected at the feed point. It immediately equilibrates with the two countercurrent fluid streams. In which direction does the solute move? Toward the condenser end, or toward the reboiler end? What type of concentration profile does the solute exhibit between the feed point and either end? If two solutes are selected that concentrate at the different ends of the extraction system, how "good" can the separation be as a function of system length? Use linear multistate chemical system theory to answer these questions.

Steady-State Multistate Convection and Reaction: The Supported Liquid-Phase Catalyst

In certain classes of homogeneous catalytic reactions that involve organometallic complexes, it is necessary to bubble reactant gases through a catalyst solution. Unfortunately, the residence time and the surface area of the gas bubbles can be inadequate, so that it is not possible to provide the mass-transfer rates needed to supply dissolved gaseous reactants at a rate that a commercial reaction requires.

If all of the reactants and products can be made to exist in the gas phase (by increasing the temperature of the system, for example), it is possible to conduct the reaction through the use of a *supported liquid-phase catalyst*, which is the catalyst analog of the liquid-containing support that is employed in gas-liquid chromatography.

The experiment is relatively simple. Fill 10–40% of the pore volume of an inert, porous solid with a nonvolatile catalyst solution in the same manner as the pore volume of the solid is filled with a nonvolatile liquid for gas chromatography applications. Pack a column with the supported liquid-phase catalyst, and run the column as a steady-state fixed-bed gas-phase reactor.

Apply the theory of multistate chemical systems to such a system, in which the reaction is assumed to be first order in a limiting gas-phase reactant i that dissolves in the liquid, where it reacts to form a product with the aid of the dissolved homogeneous organometallic catalyst. How would you propose to acquire experimental data, and how would you plot the data? What parameters would the slope contain?

Non-Steady-State Multistate Reaction: Phase-Transfer Catalysis
In *phase-transfer catalysis*, two immiscible phases are present, usually water and an immiscible organic solvent. Dissolved in both phases is a complexing agent that can transfer a reactant from the aqueous phase to the organic phase, where reaction occurs at a rate higher than in the aqueous phase. The complexing and phase transfer processes can be assumed to be equilibrium processes. Using linear multistate chemical system theory, demonstrate how phase-transfer catalysis can enhance reaction rates.

Non-Steady-State Multistate Convection: Parametric Pumping
Go to the chemical engineering literature and study the separation technique called *parametric pumping*. If possible, theoretically model the technique using linear multistate chemical system theory.

Non-Steady-State Multistate Diffusion, Convection, and Reaction: Reaction Gas Chromatography
Go to the chemistry and chemical engineering literature and learn the principle of operation of *reaction gas chromatography*. Is it possible to apply linear multistate chemical system theory to model this separation/reaction technique? If the answer is yes, do so.

References
(1) Slattery, J. C. *Momentum, Energy, and Mass Transfer in Continua;* McGraw-Hill Book Company: New York, 1972.
(2) Bird, R. B.; Stewart, W. E.; Lightfoot, E. N. *Transport Phenomena;* John Wiley & Sons: New York, 1960.
(3) MacDonald, D. D. *Transient Techniques in Electrochemistry;* Plenum: New York, 1977.
(4) Weisz, P. B. *Science* **1973**, *179* (4072), 433.
(5) de Bono, E. *CHEMTECH* **1972**, *2*, 456.
(6) Hirschfeld, T. *Energy and Technology Review;* Lawrence Radiation Laboratory: Livermore, Calif., Feb. 1984 (internal publication), 16.
(7) Rony, P. R. Virginia Polytechnic Institute and State University, unpublished course notes, 1973.
(8) Rony, P. R. In *Workshop on Microprocessors and Education, October 20–21, 1980;* IEEE Catalog No. TH0083-6; IEEE Computer Society Press: Washington, D.C., 1981; p. 67.
(9) Rony, P. R. *Sep. Sci.* **1968**, *3*, 425.
(10) *IEEE Standard Dictionary of Electrical and Electronics Terms;* ANSI/IEEE Std 100-1984, 3rd ed.; Institute of Electrical and Electronics Engineers: New York, 1984.
(11) *International Dictionary of Physics and Electronics*, 2nd ed.; D. Van Nostrand: Princeton, N.J., 1961.
(12) Prigogine, I.; Defay, R. *Chemical Thermodynamics;* Longmans, Green: London, 1954; pp. 10–13.
(13) *Comprehensive Chemical Kinetics;* Bamford, C. H.; Tipper, C.F.H., Eds.; Elsevier: Amsterdam, 1969.
(14) Atkins, P. W. *Physical Chemistry*, 3rd. ed.; W. H. Freeman: San Francisco, 1986.
(15) Geankoplis, C. J. *Mass Transport Phenomena;* Holt, Rinehart and Winston: New York, 1972.
(16) Spiegel, M. R. *Theory and Problems of Laplace Transforms;* McGraw-Hill: New York, 1965.
(17) Jenson, V. G.; Jeffreys, G. V. *Mathematical Methods in Chemical Engineering*, 2nd ed.; Academic: New York, 1977.
(18) Spiegel, M. R. *Advanced Mathematics for Engineers and Scientists;* McGraw-Hill: New York, 1971.
(19) Spiegel, M. R. *Mathematical Handbook of Formulas and Tables;* McGraw-Hill: New York, 1968.
(20) Crank, J. *The Mathematics of Diffusion;* Oxford University Press: London, 1957.
(21) Welty, J. R.; Wicks, C. E.; Wilson, R. E. *Fundamentals of Momentum, Heat, and Mass Transfer*, 2nd ed.; John Wiley and Sons: New York, 1976.
(22) Thiele, E. W. *Ind. Eng. Chem.* **1939**, *31*, 916.
(23) Giddings, C. *Sep. Sci.* **1966**, *1*, 123; **1967**, *2*, 797.
(24) Rony, P. R. *Sep. Sci.* **1969**, *4*, 413.
(25) Rony, P. R. *Sep. Sci.* **1969**, *4*, 447.
(26) Prausnitz, J. M. *The Thermodynamic Theory of Phase Equilibria* (lecture notes); University of California: Berkeley, 1961.
(27) Rony, P. R. *A General Approach to Chemical Separations;* Monsanto Corporation: St. Louis, 1967.

NOTE: This chapter contains text, figures, and equations from an unpublished manuscript and set of lecture slides, copyright © by Peter R. Rony 1973 and 1987 (all rights reserved) for an elective course entitled *Chemical Microengineering*. Readers are encouraged to contact the author for information on this material, which contains both a complete derivation of the continuity-of-species equation for linear multistate chemical systems and a brief discussion of the concept of a dimensionless group and its application in such systems.

Chapter 5
BUILDING ENTHUSIASM FOR QUANTUM MECHANICS AND STATISTICAL MECHANICS

HENRY A. MCGEE, JR.
Professor of Chemical Engineering
Virginia Polytechnic Institute and State University
Blacksburg, Va. 24061

Whether in pure or applied chemistry, one must have a molecular point of view to complement one's equally essential macroscopic or continuum perspective. This molecular point of view may only be developed from a study of quantum mechanics and statistical mechanics, but no subjects in physical chemistry are more esoteric and abstract than these two. Students are faced, for example, with the perplexing idea of wave-particle duality, and with the demand that they replace the perfectly acceptable macroscopic concept of momentum with an operator, and even that, an imaginary-looking thing, $(h/2\pi i)\partial/\partial x$. Even when they have mastered the complex mathematics, students are left with strange sorts of insights such as tunneling and diffuseness of location. In statistical mechanics, we begin by wondering about the number of identifiable ways one might arrange unlabelable objects into separate containers or groups. It is all so unreal.

But the mystical character of these subjects may be erased and, simultaneously, enthusiasm may be built for the difficult task of learning these ideas and their associated complex mathematics when the students realize and are frequently reminded of the utility and value of the theory in the real world. Both of these subject areas are immediately useful to pure and applied chemists in understanding the equilibrium or thermodynamic properties of matter. Of course, quantum and statistical mechanics can always suggest what variable one might plot against what in order to gain maximum insight into phenomena of interest. But more directly, quantum and statistical mechanics also allow the actual evaluation of thermodynamic properties to numerical accuracies that are sufficient for many applications in chemistry. Every student is familiar with C_p^0 as a simple polynomial in temperature that is readily useful in calculations of compression and expansion, the effect of temperature upon a reaction equilibrium, and so on. Every student is also familiar with the use of absolute entropies of ideal gases in calculations of chemical reaction equilibria in real gas phases. What most students do not recognize is that such numbers have been largely deduced from statistical mechanics and quantum mechanics. The more precisely one performs a calorimetric experiment, the more nearly will one check the prediction from the theory.

Statistical mechanics and thermodynamic properties

It is strikingly impressive that *all* of the thermodynamic properties may be expressed so neatly and concisely in terms of the partition function,

$$f = \sum_i e^{-\epsilon_i/kT}$$

For example, consider the internal energy, which is defined as

$$U \equiv \sum_i n_i \epsilon_i$$

This chapter was excerpted from a textbook, *Molecular Engineering*, by Henry McGee, now in draft form. All rights reserved.

and then
$$U = \sum_i n e^{-\epsilon_i/kT} \epsilon_i / f$$

The derivative of f with temperature is
$$\left(\frac{\partial f}{\partial T}\right)_V = \frac{1}{kT^2} \sum_i \epsilon_i e^{-\epsilon_i/kT}$$

where the volume is held fixed because the translational energy levels depend upon volume, as we shall see. Then
$$U = nkT^2 \left(\frac{\partial \ln f}{\partial T}\right)_V$$

and similar relationships may be developed for all thermodynamic properties. These are summarized for real gases and plasmas, not solely ideal gases, in the box. The partition function approach is completely general and applicable to all states of matter. Approximation or uncertainty comes in the specification of energy levels in specific sorts of molecular circumstances and in the evaluation of f itself.

All of the statistical thermodynamic arguments developed herein utilize single particle partition functions. Interacting systems, such as real gases, require formal arguments beyond those presented here.

Box 1. Summary of all thermodynamic properties of fluids in terms of the partition function. (Somewhat different relationships may be derived for solids.)

$$U = nkT^2 \left(\frac{\partial \ln f}{\partial T}\right)_V$$

$$C_v = \left(\frac{\partial U}{\partial T}\right)_V$$

$$S = nk\left(\ln \frac{f}{n} + 1\right) + nkT \left(\frac{\partial \ln f}{\partial T}\right)_V$$

$$A = -nkT\left(\ln \frac{f}{n} + 1\right)$$

$$p = nkT \left(\frac{\partial \ln f}{\partial V}\right)_T$$

$$H = nkT^2 \left(\frac{\partial \ln f}{\partial T}\right)_V + nkTV \left(\frac{\partial \ln f}{\partial V}\right)_T$$

$$C_p = \left(\frac{\partial H}{\partial T}\right)_p$$

$$G = -nkT\left(\ln \frac{f}{n} + 1\right) + nkTV \left(\frac{\partial \ln f}{\partial V}\right)_T$$

Consider first an ideal gas in which the energy content of a molecule is separable into translational, rotational, vibrational, and electronic contributions,
$$\epsilon_{\text{total}} = \epsilon_{\text{trans}} + \epsilon_{\text{rot}} + \epsilon_{\text{vib}} + \epsilon_{\text{elect}}$$

Such separation is a good approximation that immediately leads to the partition function, f, being expressed as
$$f = f_{\text{trans}} \times f_{\text{rot}} \times f_{\text{vib}} \times f_{\text{elect}}$$

Separability fails somewhat for vibrational and rotational modes, but even here, calculations that include such interactions reveal little resulting impact on the thermodynamic properties. In addition, data to characterize such interactions are rare.

Monatomic species

The energy levels of a monatomic species come from the quantum mechanical problem of a particle in a one-dimensional box for which we obtain energies,
$$\epsilon = h^2 n^2 / 8ml^2 \quad \text{where } n = 1, 2, 3, \ldots$$

Here n is the quantum number and l is the length of the line, that is, the volume of the one-dimensional box in which the particle moves. The energy levels are the same in all three dimensions and are sufficiently close together that integration rather than summation of the exponential terms is an excellent approximation, and one obtains finally

$$f_{\text{trans}} = \left(\frac{2\pi mkT}{h^2}\right)^{3/2} V$$

The only additional energy that the ideal monatomic species may contain is that due to electronic excitation. This partition function is written

$$f_{\text{elect}} = g_0 + g_1 e^{-\epsilon_1/kT} + g_2 e^{-\epsilon_2/kT} + \ldots$$

where the ground electronic state, ϵ_0, has been assigned zero energy, and where the upper energy levels are given by ϵ_1, ϵ_2, etc. The multiplicity of each electronic state is given by the integers g_0, g_1, etc. The hydrogen atom spectrum shown in Figure 1 is similar to that of all species. The multiplicities are all equal to 2 in this case, and we

Figure 1. Electronic energy levels of the hydrogen atom.

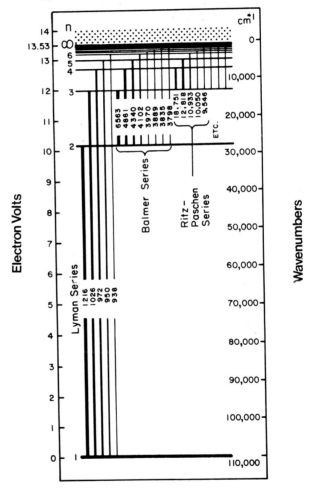

note that the first excited level is approximately 10 eV above the ground state, and the species ionizes at 13.6 eV. Ten eV is equivalent to approximately 80,000 cm^{-1}, and even at 1,000 K, kT is 1.4×10^{-20} J or 700 cm^{-1}. Even very high temperatures fail to electronically excite atoms (and molecules), and the electronic partition function remains at g_0. As another example, lithium has a low-lying ionization potential and a low-lying first excited level at only 1.84 eV. Even here, the first excited level is only 1% populated even at 4500 K.

The complete partition function, then, for all monatomic species up to temperatures insufficiently high to excite electronic levels is

$$f_{\text{mono}} = g_0 \left(\frac{2\pi mkT}{h^2}\right)^{3/2} V \qquad (1)$$

where m is the atomic mass and all other symbols have their usual meaning. For all monatomic species at other than very high temperatures,

$$U^0 - U_0^0 = nkT^2 \left(\frac{\partial \ln f}{\partial T}\right)_V = \frac{3}{2} RT.$$

The heat capacity is independent of temperature and is given by

$$C_v^0 = 3/2 \, R \text{ and } C_p^0 = \frac{5}{2} R$$

The entropy is given by (see Box 1, page 68)

$$S^0 = nk \left[\ln(f/n) + 1 + T \left(\frac{\partial \ln f}{\partial T}\right)_V \right]$$

which in practical units and in dimensionless form becomes

$$S^0/R = 3/2 \ln M + 5/2 \ln T + \ln g_0 - \ln p + 3.4533 \qquad (2)$$

where M is the molecular weight in grams, T the absolute temperature in kelvins, and p the pressure in kPa. This is the so-called Sackur-Tetrode equation. A comparison between such calculated entropies and experimentally determined entropies appears in Table I.

Table I. Calculated vs. experimental entropies for several monatomic species at 298.15 K

Species	S^0(calc)	S^0(exp)
He	30.11	30.4
Na	36.70	37.2
Hg	41.78	42.2
Ar	36.98	36.85

Source: Reference 2.

Recall that the experimental third-law entropy is given by

$$S^0(T,p) = \int_0^{T_f} \frac{C_p(\text{sol})}{T} dT + \frac{\Delta H_f}{T_f} + \int_{T_f}^{T_b} \frac{C_p(\text{liq})}{T} dT + \frac{\Delta H_v}{T_b} + \int_{T_b}^{T} \frac{C_p(\text{gas})}{T} dT + (S - S^0)_{T,p=1} + \int_1^p R \ln p$$

where all of the indicated data—heat capacities as a function of temperature for all phases, heat of fusion and melting point, heat of vaporization and boiling point—must be calorimetrically determined, and we even need an equation of state to evaluate the non-ideality term, $(S - S^0)$. Contrast this requirement for data with that from the molecular perspective of only T, p, g_0, and the atomic mass. Very uncharacteristically in chemistry, the theoretical result is here *more* accurate than the experimental result.

Practice Problem:

Verify the theoretically calculated absolute entropies in Table I.

The properties of any monatomic species may be readily evaluated as a function of temperature. Results are displayed in Table II for some typical species.

Table II. Thermodynamic properties of some monatomic species at 1 atm

	H			Ar	Li	Hg	C
T	C_p^0/R	$(H^0 - H_0^0)/RT$	S^0/R	S^0/R	S^0/R	S^0/R	S^0/R
100	2.500	2.500	11.054	15.880	13.952	13.294	—
200	2.500	2.500	12.789	17.613	15.684	20.026	—
300	2.500	2.500	13.800	18.627	16.694	21.036	19.015
400	2.500	2.500	14.519	19.346	17.416	21.758	19.737
600	2.500	2.500	15.533	20.360	18.426	22.768	20.747
800	2.500	2.500	16.252	21.079	—	—	21.469
1000	2.500	2.500	16.810	21.637	19.701	—	22.034

Source: Reference 3.

Note that in Table II the heat capacity and reduced enthalpy of all species are identical and are also independent of temperature. The entropy, however, is a function of both species and temperature.

Practice Problem:
Verify the numbers in Table II.

Classical thermodynamics yields only interrelationships. There is no sense of the values of properties and certainly no sense of why the observable properties have the values they do. All of this real-world insight from statistical thermodynamics has come from considering a particle in a box from the perspective of the other-worldly Schrodinger wave equation. Note that we are also able to calculate the properties of species like H and C as monatomic ideal gases. Such species appear in, for example, high-temperature processes. Their properties are important, but calorimetric measurements are impossible.

Linear molecules

Linear molecules may absorb energy in rotation and in vibration as well as in translation and electronic excitation, as was the case with monatomic species. The quantum mechanical problem of the rigid rotator is readily solved to yield energy levels,

$$\epsilon_{rot} = (J+1)Jh^2/8\pi^2 I \quad \text{where } J = 0, 1, 2, 3 \ldots$$

where J is the rotational quantum number and I is the moment of inertia about the center of mass. With h in J s and I in kg m^2, ϵ_{rot} is in joules.

If we idealize the molecular vibrations as being harmonic, the energy levels obtained from solving the appropriate Schrödinger equation are

$$\epsilon_{vib} = (n + 1/2)h\nu = (n + 1/2)hc\,\omega \quad \text{where } n = 0, 1, 2, 3 \ldots$$

where n is the vibrational quantum number, and ν is the fixed harmonic frequency in s^{-1}. Frequencies are, however, more frequently specified in wave numbers, ω, with units of cm^{-1}, using c, the velocity of light, in cm s^{-1}. Note that the energy levels are all equally spaced and that transitions between adjacent levels are all of energy $h\nu$.

A schematic representation of experimental electronic, rotational, and vibrational energy levels appears in Figure 2, where the real anharmonicity of vibration results in unequally spaced energy levels.

Figure 2. Schematic representation of electronic, vibrational, and rotational energy levels.

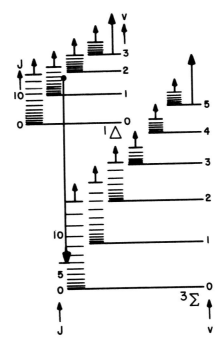

The rotational energy levels are typically sufficiently closely spaced that, again, integration rather than summation is adequate for the evaluation of the partition function. One needs only the moment of inertia or its equivalent, the rotational constant, to evaluate the rotational partition function. The rotational constant is defined as $B(\text{cm}^{-1}) = (h/8\pi^2 Ic) 10^{-2}$, where c is the velocity of light in cm s^{-1}, or as $B(\text{MHz}) = (h/8\pi^2 I) 10^{-6}$. So to know either $I(\text{kg m}^2)$ or $B(\text{cm}^{-1})$ or $B(\text{MHz})$ is to know them all. Some typical rotational data on representative linear molecules appear in Table III. Note the extraordinary precision with which rotational constants may be measured. Also note that a linear molecule has the same moment about each of its two axes of rotation perpendicular to the molecular line of centers. Hence, only one I or B appears in Table III.

Omitting the mathematical exercise, suffice it to say that the rotational partition function becomes

$$f_{\text{rot}} = 8\pi^2 IkT/\sigma h^2$$

The quantity σ is called the symmetry number and is available by inspection of the molecule. It is equal to 1 for asymmetric molecules and 2 for symmetric molecules. It arises in the partition function because only even or odd quantum numbers, J, are allowed for symmetric linear molecules.

The partition function for a single vibrator is

$$f_{\text{vib}} = e^{-h\nu/2kT} + e^{-3h\nu/2kT} + e^{-5h\nu/2kT} + \ldots$$

Arbitrarily setting the zero of energy at $n = 0$ (rather than its quantum mechanical value of $h\nu/2$), and recognizing that Σa^n where $a < 1$ is $(1 - a)^{-1}$, we can write

$$f_{\text{vib}} = (1 - e^{-h\nu/kT})^{-1}$$

Table III. Rotational data on several linear molecules (moments are all of order 10^{-47})

Molecule	B (MHz) or I (kg m^2)	Molecule	B (MHz) or I (kg m^2)
H$_2$	0.473×10^{-47}	CS	24,495.592
F$_2$	31.8×10^{-47}	Al^{35}Cl	7,288.73
Cl$_2$	113.5×10^{-47}	COS	6081.49255
Br$_2$	341.4×10^{-47}	HCN	44,315.9757
I$_2$	740.5×10^{-47}	FCN	10,544.20
F^{35}Cl	15,483.688	LiOH	35,342.44
F^{37}Cl	15,189.221	N$_2$O	$\{66.1 \times 10^{-47}$ / $12,561.6338\}$
HI	192,658.8	HC≡C—C≡N	4,549.067
O$_2$	19.2×10^{-47}	C$_2$N$_2$	175×10^{-47}
N$_2$	13.8×10^{-47}	C$_3$O$_2$	395.2×10^{-47}
CO	57,635.970	OH	1.513×10^{-47}
H^{35}Cl	312,991.30		

Source: Reference 3.

The scale change wherein we arbitrarily set the energy of the vibrator in its ground state ($n = 0$) equal to zero is of no consequence, as absolute energy has no meaning anyway. The zero of energy is an arbitrary assignment of convenience. The omitted factor in the partition function would not have appeared in any event in $C_v^0(\text{vib})$, nor would it appear at all in the expression for the entropy. It would appear as an additive constant in all four of the energy functions, but only differences in $U, H, A,$ or G have meaning, so it again cancels out. To remind us of this arbitrary choice of energy zero we will always write

$(H^0 - H_0^0)$, $(G^0 - G_0^0)$, etc.

The total partition function, then, for all diatomic molecules is given by

$$f_{\text{diatomic}} = g_0 \left(\frac{2\pi mkT}{h^2}\right)^{3/2} V \frac{8\pi IkT}{\sigma h^2} (1 - e^{-h\nu/kT})^{-1} \quad (3)$$

and the thermodynamic properties of all diatomics as ideal gases are readily calculated from this partition function. Data on several typical diatomic species from which the thermodynamic properties may be calculated appear in Table IV. In this table, the ground state of each species is described by a spectroscopic term symbol, the exact meaning of which need not concern us here. The superscript number to the left of the Greek letter is the multiplicity of the electronic ground state of the molecule, g_0, which, of course, determines the electronic partition function at other than extraordinarily high temperatures. The multiplicity, g_0, is unity for all species other than free radicals. A few stable molecules like O_2 have other than singlet ground states; thus g_0 for these species is not unity, and they are technically free radicals.

Table IV. Spectroscopically measured properties of some typical diatomic molecules

Molecule	State	$\omega(\text{cm}^{-1})$	$B(\text{MHz})$
H_2	$^1\Sigma_g^+$	4395	1,774,200
F_2	$^1\Sigma_g^+$	892	26,390
Cl_2	$^1\Sigma_g^+$	565	7,895
Br_2	$^1\Sigma_g^+$	323	2,458
I_2	$^1\Sigma_g^+$	215	1,133
OH	$^2\pi$ [a]	3735	554,660
O_2	$^3\Sigma_g^-$	1580	43,708
N_2	$^1\Sigma_g^+$	2360	60,812
CO	$^1\Sigma^+$ [a]	2170	58,612

The partition function for all diatomic molecules may be inserted in any of the relationships that provide the thermodynamic properties. For example, the entropy may be expressed as

$$\frac{S^0(T,p)}{R} = \left\{\ln\left[g_0\left(\frac{2\pi mkT}{h^2}\right)^{3/2}\frac{kT}{p}\right] + \frac{5}{2}\right\} + \left\{\ln\frac{8\pi IkT}{\sigma h^2} + 1\right\} + \left\{\frac{h\nu}{kT}(e^{h\nu/kT} - 1)^{-1} - \ln(1 - e^{-h\nu/kT})\right\}$$

which in practical units becomes

$$\frac{S^0(T,p)}{R} = \left[\frac{3}{2}\ln M + \frac{5}{2}\ln T + \ln g_0 - \ln p + 3.4533\right] + [\ln T - \ln \sigma - \ln B + 10.9444] + S^0(\text{vib})/R \tag{4}$$

where M is the molecular weight in grams, p is in kPa, B in MHz, T in kelvins, and the last term, because it is somewhat more tedious to evaluate, appears in the table of so-called Einstein functions. It arises, of course, from the harmonic vibrational contribution to the entropy. This vibrational contribution is also readily coded for machine computation. Einstein first developed the harmonic vibrator idea to describe the then inexplicable variation of the heat capacity of solids with temperature. Although its success was remarkable, it failed at very low temperatures because atoms or molecules do not vibrate about their equilibrium lattice positions with a single frequency, as Einstein had assumed. Einstein's theory is now used solely to describe intramolecular vibrational modes.

The heat capacity can be similarly developed, and we find that that important quantity for all diatomic molecules is given by

$$C_p^0 = \frac{3}{2}R + \frac{2}{2}R + C_v^0(\text{vib}) + (C_p^0 - C_v^0)$$

where

$$\frac{C_v^0(\text{vib})}{R} = \left(\frac{h\nu}{kT}\right)^2 e^{h\nu/kT}/(e^{h\nu/kT} - 1)^2 \tag{5}$$

where again the $C_v^0(\text{vib})$ is similarly read from the table of Einstein functions. Note that all of the temperature dependence of the heat capacity arises from vibration. The quantum oscillator makes sense of the phenomenological parameter of heat capacity.

Some comparisons of experimental heat capacities and entropies with theoretical values for several diatomics at several temperatures appear in Table V. The data agree well.

Table V. Calculated and experimental heat capacities and entropies of several diatomic species at 101.325 kPa (1 atm)

Species	T(K)	Experiment		Theory	
		C_p^0/R	S^0/R	C_p^0/R	S^0/R
O_2	100	3.546	21.736	3.501	20.835
	200	3.510	24.176	3.503	23.262
	300	3.537	25.602	3.534	24.687
	400	3.622	26.630	3.621	25.714
N_2	100	3.613	19.171	3.500	19.204
	200	3.515	21.625	3.501	21.631
	300	3.508	23.048	3.503	23.051
	400	3.521	24.059	3.581	24.060
CO	300	3.549	23.560	3.505	23.782
	400	3.539	24.801	3.529	24.792
	500	3.553	24.976	3.583	25.585

Practice Problem:
Verify any number in the theoretical column of Table V.

An Example:
High-powered lasers such as the high-frequency (HF) device that radiates on several lines near 2.7 µm have been touted as possible futuristic weapon systems. To produce such a device requires that one first produce a population inversion of the HF vibrators. This is readily done on a large scale by using the fast reaction

$$F + H_2 \rightarrow HF^* + H$$

in a supersonic flow system that yields product HF initially in the $\nu = 1, 2,$ and 3 levels and in amounts far in excess of the occupation numbers allowed by the Boltzmann distribution,

$$n_i = n\, e^{-\epsilon_i kT}/f.$$

The fast reaction produces a nonthermodynamic product mix with a highly inverted population. The vibrational energy is stimulated out of the molecule in an optical cavity to produce the giant laser beam. Simultaneously the vibrators come to equilibrium, that is, a Boltzmann distribution, by V-V collisional exchange that is faster than the relaxation of vibrational energy into the rotation/translation heat bath. At this equilibrium, the vibrators are characterized by a vibrational temperature that is well in excess of the rotation/translation temperature. This latter temperature could be measured with a thermocouple. Such a now thermodynamic vibrational system also signals the end of any further laser action. The equilibrium vibrators then more slowly leak their energy into the rotation/translation heat bath, and, when the process is complete, the entire system becomes characterized by a single parameter called the adiabatic flame temperature.

Linear molecules

Not just diatomic, but all linear molecules are handled similarly, and each vibrational contribution is considered as completely separable from all the rest. Since there are 3 degrees of translational freedom and 2 degrees of rotational freedom, there will always be $3n - 5$ vibrational frequencies for an n-atom linear molecule. For example, dicyanoacetylene has 6 atoms and 13 Einstein vibrators that contribute to all of the thermodynamic properties. The partition function for any linear molecule regardless of its complexity is given by

$$f_{\text{lin}} = g_0 \left(\frac{2\pi mkT}{h^2}\right)^{3/2} V \frac{8\pi^2 IkT}{\sigma h^2} \prod_{i=1}^{i=3n-5} (1 - e^{-h\nu_i/(kT)})^{-1} \quad (6)$$

and all of the thermodynamic properties may be developed from this function.

It is striking that all of this real-world insight into why the properties have the values that they do has come from the other-worldly quantum mechanical problems of the rigid rotator and the harmonic oscillator.

Nonlinear molecules

Nonlinear rigid molecules differ only in having an additional degree of rotational freedom, for the mass of the molecule is now distributed about three principal axes of rotation rather than only two. The quantum mechanical problem of a three-dimensional rotator is readily solved, the energy levels are sufficiently close together that integration rather than summation is again adequate, and the rotational partition function may be expressed as

$$f_{\rm rot} = \frac{\pi^{1/2}(8\pi^2 kT)^{3/2}(I_A I_B I_C)^{1/2}}{\sigma h^3}$$

where I_A, I_B, and I_C are the principal moments of inertia in kg m^2. The symmetry number, σ, has the same meaning for nonlinear molecules as it did for linear molecules, and its value can be determined by inspection as the total number of indistinguishable orientations of the molecule. For example, CH_3Cl can be rotated 120° three times about the C—Cl bond with no observable change. Thus σ for CH_3Cl is 3. Planar benzene can be rotated six times about an axis perpendicular to the plane and then turned over and rotated six more times. Thus σ for C_6H_6 is 12. And so on. Table VI lists symmetry numbers for several species. Note that the molar entropy varies as $-R \ln \sigma$. One would expect, for example, that the entropy of monodeuterobenzene would be more than that of benzene by $-R(\ln 2 - \ln 12)$ or 7.5 J/mol K.

Table VI. Symmetry numbers of several molecules

Cpd	σ	Cpd	σ
C_2H_2	2	i-C_4H_{10}	3
CO_2	2	n-C_4H_{10}	2
CO	1	C_2H_6	2
H_2	2	CCl_4	12
CH_4	12	C_2H_4	4
CH_3Cl	3	UF_6	24
$(CH_3)_2CO$	2	NH_3	3
C_6H_6	12	SF_6	24
BF_3 (planar)	6	*trans* cy-C_6H_{12} (chair form)	6
		cis cy-C_6H_{12} (boat form)	2
H_2O	2	C_3H_8	2
		BF_4^- (tetrahedral)	4

Table VII. Rotational data on several representative nonlinear molecules (each moment is of order 10^{-47})

Molecule	B(MHz) or I(kg m^2)	Molecule	B(MHz) or I(kg m^2)
HC—CH ‖ ‖ HC CH \\NH/	9130.610 9001.343 4532.083	NO_2	3.486×10^{-47} 63.76×10^{-47} 67.25×10^{-47}
HC—CH ‖ ‖ HC CH \\O/	9447.153 9246.775 4670.846	CH_4 SiH_4	5.47×10^{-47} (same moment about each axis) 9.78×10^{-47} (same moment about each axis)
$(CH_3)_2NH$	34242.22 9334.03 8215.98	C_3H_4	98.276 98.276 5.499
$H_2C{=}CH^{35}Cl$	56840.21 6029.96 5445.29	*cis*-C_4H_8	59.977 142.431 191.886
CH_2O	38835.369 34003.282	*trans*-C_4H_8	25.675 224.40 239.556
SO_2	60778.516 (13.3×10^{-47}) 10317.963 (86.2×10^{-47}) 8799.808 (99.4×10^{-47})	cy-C_6H_{12} NH_3	$I_1 \times I_2 \times I_3 = 12.583 \times 10^{-138}$ 2.782×10^{-47} 2.782×10^{-47} 4.33×10^{-47}

Practice Problem:
Verify any number in Table VI.

Some typical values for moments, products of moments, or rotational constants appear in Table VII.

Because there are now three degrees of rotational freedom, there will be $3n - 6$ vibrational contributions to the properties. The partition function for all nonlinear rigid molecules is given by

$$f = g_0 \left(\frac{2\pi m k T}{h^2}\right)^{3/2} V \frac{\pi^{1/2}(8\pi^2 k T)^{3/2}(I_A I_B I_C)^{1/2}}{\sigma h^3} \prod_{i=1}^{i=3n-6}(1 - e^{-h\nu_i/kT})^{-1}$$

$$\underbrace{}_{\text{electronic part}} \quad \underbrace{}_{\text{translational part}} \quad \underbrace{}_{\text{rotational part}} \quad \underbrace{}_{\text{vibrational part}} \quad (7)$$

The logarithm of this partition function is given by

$$\ln f = \frac{3}{2} \ln T + \frac{3}{2} \ln T - \sum_{i=1}^{i=3n-6} \ln(1 - e^{-h\nu_i/kT}) + \ldots$$

Its derivative with respect to temperature is given by

$$\left(\frac{\delta \ln f}{\delta T}\right) = \frac{3}{2}\left(\frac{1}{T}\right) + \frac{3}{2}\left(\frac{1}{T}\right) + \sum_{i=1}^{i=3n-6} \frac{e^{-h\nu_i/kT}\left(\frac{h\nu_i}{kT^2}\right)}{(1 - e^{-h\nu_i/kT})}$$

and, for example, the enthalpy function is given by

$$\frac{H^0 - H_0^0}{RT} = 4 + \sum_{i=1}^{3n-6} \frac{h\nu_i}{kT}(e^{h\nu_i/kT} - 1)^{-1}$$

The temperature variation of enthalpy arises wholly from the vibrational modes. The contribution to the enthalpy due to each vibrator is read from the Einstein table. This Einstein enthalpy function is also readily coded for machine computation.

As an example, consider the calculation of $(H^0 - H_0^0)/RT$ for NH_3 at 400 K. NH_3 is a pyramidal molecule with $(3n - 6)$ or six frequencies. Because of symmetry, there are two frequencies that appear separately (so-called nondegenerate frequencies) at 3337 and 950 cm^{-1}. There are also two frequencies that are repeated once each (so-called doubly degenerate frequencies) at 3414 and 1628 cm^{-1}.

ω(cm^{-1})	ω/T	$(H^0 - H_0^0)/RT$
3337	8.343	—
950	2.375	0.116 (1)
3414	8.535	—
1628	4.070	0.017 (2)
	Total	0.15

The complete enthalpy is then

$$(H^0 - H_0^0)/RT = 4 + 0.15 = 4.15$$

The entropy, which depends upon $\ln(f)$ as well as upon the derivative of $\ln(f)$ may be written in practical units as

$$\frac{S^0_{(\text{elect})}}{R} = \ln g_0$$

$$\frac{S^0_{(\text{trans})}}{R} = \frac{3}{2}\ln M + \frac{5}{2}\ln T - \ln p + 3.4533$$

$$\frac{S^0_{(\text{rot})}}{R} = \frac{3}{2}\ln T - \frac{1}{2}\ln B_x B_y B_z - \ln \sigma + 16.9890$$

$$\frac{S^0_{(\text{vib})}}{R} = \sum_i \frac{h\nu i}{kT}(e^{h\nu_i/kT} - 1)^{-1} - \ln(1 - e^{-h\nu_i/kT}) \quad (8)$$

Practice Problem:
Verify the several components of the entropy of nonlinear rigid molecules as given by Equation 8. In particular, verify the constants that appear in the expressions for the translational and the rotational entropy.

The expression for the important Gibbs free energy function may be similarly developed:

$$\left(\frac{G^0 - H_0^0}{RT}\right)_{\text{elect}} = -\ln g_0$$

$$\left(\frac{G^0 - H_0^0}{RT}\right)_{\text{trans}} = \frac{5}{2} - \frac{3}{2}\ln M - \frac{5}{2}\ln T + \ln p - 3.4533$$

$$\left(\frac{G^0 - H_0^0}{RT}\right)_{\text{rot}} = \frac{3}{2} - \frac{3}{2}\ln T + \frac{1}{2}\ln B_x B_y B_z + \ln \sigma - 16.9890$$

$$\left(\frac{G^0 - H_0^0}{RT}\right)_{\text{vib}} = \sum_{i=1}^{i=3n-6} \ln(1 - e^{-h\nu_i/kT}) \tag{9}$$

Practice Problem:
Verify the component expression appearing in Equation 9.

With this ability to calculate Gibbs free energies, we can deduce equilibrium constants for real gas reactions:

$$\Delta G^0(T, p=1) = RT \sum \left(\frac{G^0 - H_0^0}{RT}\right) + \Delta H_0^0 \tag{10}$$

Here, ΔH_0^0 is the heat of reaction at 0 K. We thus gain insight into why the equilibrium constant for some reactions is large while for others it is small. And we observe the effect of symmetry or of moment of inertia upon the equilibrium constant.

As another example problem, let us calculate C_p^0 of ethylene at 464.00 K. The calorimetric value at this temperature is 59.208 J/mol K. The molecular weight is 28.05, the symmetry number is obviously 4, the rotational constants are 4.86596, 1.001329, and 0.828424 cm^{-1}, and the vibrational frequencies in cm^{-1} are 3026, 1623, 1342, 1023, 3103, 1236, 949, 943, 3106, 826, 2989, and 1444. Note that there are 12 frequencies, as we know must be the case. The heat capacity is given by

$$C_p^0 = \frac{3}{2}R + \frac{3}{2}R + \sum_{i=1}^{12} C_v^0(\text{vib}) + R$$

wherein the vibrational contributions are immediately developed from the Einstein table.

ω/T	$C_v^0(\text{vib})/R$	ω/T	$C_v^0(\text{vib})/R$
6.694	0.00632	2.892	0.27857
6.688	0.00636	2.664	0.33264
6.522	0.00745	2.205	0.45948
6.442	0.00825	2.045	0.50885
3.498	0.16735	2.032	0.51295
3.112	0.23308	1.780	0.59477

Here, $\sum C_v^0(\text{vib})/R = 3.11607$. With the vibrational contributions in hand, we immediately calculate $C_p^0 = 59.166$ J/mol K, which compares well with the calorimetric value.

It is worthwhile to note in passing that these values of C_p^0, together with similar data on H$_2$ and on solid graphite, enable us to calculate ΔH_f^0 and ΔG_f^0 of ethylene at any temperature provided only that we have a measured heat of formation at some one temperature, as is apparent from Equation 10. The heat of formation cannot be deduced from statistical thermodynamics, but it can, in principle, be calculated quantum mechanically; we will subsequently discuss a particularly useful technique for doing just this.

Practice Problem:

Calculate the entropy of ethylene as an ideal gas at its normal boiling point of 169.40 K. The experimental value has been determined as follows (4):

0 – 15 K	Debye extrapolation (theoretical)	1.046
15 – 103.95 K	graphical integration	51.154
800.8/103.95	fusion	32.234
103.95 – 169.40	graphical integration	33.154
3237/169.40	vaporization	79.956
	entropy of real gas	197.54
	correction to ideal	0.63
	ideal entropy	198.17 J/mol K

Practice Problem:

Calculate the entropy of benzene as an ideal gas at 298.15 K. The calorimetric value is 269.70 J/mol K. As a symmetric planar molecule, benzene has the same moment of 147.63×10^{-47} kg m^2 about two axes and twice that, or 295.26×10^{-47} kg m^2, about the third axis. The molecule has 10 nondegenerate frequencies and 10 doubly degenerate frequencies of:

single:			double:		
3062	1008		849	1596
992	1520		3045	1178
1190	538		1485	606
672	1854		1037	1160
3063	1145		3047	404

Nonrigid molecules

Groups within a molecule may rotate about a single bond that joins that group to the remainder of the molecule. Consider, for example, methanol, where the hydroxyl hydrogen is not in line with the C—O bond. Internal rotation of the —OH can occur, and the rotation is hindered as the hydroxyl hydrogen passes by each of the three hydrogens of the methyl group. Depending upon the molecule, the internal rotation may be free or hindered or even changed in character to a low-frequency torsional oscillation as the barrier to rotation increases from zero to a few kJ to very large. In dimethyl cadmium, the methyls are so far apart that they do not experience internal retardation, and free rotation occurs. In methanol, hindered rotation occurs. In 1,3-butadiene, the barrier to internal rotation is so high that two structural isomers, cis and trans, can exist.

For free rotation, the partition function is given by,

$$f_{\text{int rot}} = \frac{2.7928}{n} (10^{45} I_r T)^{1/2}$$

where n is a sort of internal symmetry number equal to the number of equivalent orientations of the rotating group relative to the rest of the molecule. In Cd(CH$_3$)$_2$, $n = 3$. I_r is the internal moment of inertia in kg m^2, and T is the temperature in Kelvins. Since $f_{\text{int rot}}$ varies as $T^{1/2}$, free internal rotation always contributes a constant $\frac{1}{2}$R to the heat capacity. On the other hand, a high barrier, as in ethylene, results in a low-frequency torsion whose contribution to the heat capacity is determined as an Einstein oscillator which, in the usual way, contributes between 0 and R to the heat capacity. Some typical torsional frequencies appear in Table VIII. In this table, multiple groups are taken to be oscillating independently, as are all other Einstein oscillators.

Table VIII. Some typical torsional frequencies

Molecule	Frequency (cm^{-1})
CH$_3$CHO	150
CH$_3$CHF$_2$	222
(CH$_3$)$_2$CO	109
(CH$_3$)$_2$O	242
(CH$_3$)$_3$N	269

Source: Reference 5.

If rotation, either free or hindered, can occur, there will be $3n - 6 - r$ oscillators, where r is the number of such rotating groups within the molecule. For example, acetone has two rotating methyls, $r = 2$, and there are 22 vibrational frequencies.

The entropy of $Cd(CH_3)_2$ can be summarized as shown in Table IX. The result agrees well with the calorimetric third law entropy of 302.92 J/mol K.

Table IX. Entropy of dimethyl cadmium at 298.15 K

Translation and rotation	253.80 J/mol K
Vibration ($3n - 7$ frequencies)	36.65
Free internal rotation	12.26
Total	302.71 J/mol K

Source: Reference 6.

The contribution of hindered rotation to any thermodynamic property can be calculated if we know the internal moment of inertia, I_r, and the barrier height. Both of these experimental data must be supplied by the spectroscopist, who also supplies the vibration frequencies and the moments of inertia of the molecule as a whole. Some typical experimental barrier heights appear in Table X along with some quantum mechanically calculated values (explained later). Understanding the contributions to the thermodynamic properties from hindered rotators is somewhat tedious, and is outside the scope of the present discussion.

Table X. Some typical barriers to free internal rotation. Theoretical values are in parentheses.

Compound	Barrier (kJ/mol)
CH_3CH_3	12.13 (4.18)
CH_3CF_3	14.56
CH_3NH_2	8.37 (4.60)
BH_3PF_3	13.56
styrene	9.20
CH_3OH	4.60 (2.93)

Source: Reference 7.

Two Examples:

There once was high interest in the so-called electrothermal arc-jet thruster as an engine for the propulsion of spacecraft or for positioning satellites in space. The thermodynamic efficiency of any jet thruster varies approximately as (flame temperature/molecular weight)$^{1/2}$. So one wants the highest possible temperature and the lowest possible molecular weight of the working fluid entering the standard converging/diverging rocket nozzle. The maximum temperature of a chemical flame is limited by heats of chemical reactions. But an arc using electricity from solar cells or fuel cells could heat a low molecular weight fluid like H_2 to much higher temperatures than would ever be chemically possible. To study the efficacy of such a device, we must have the properties of such very hot hydrogen. The properties cannot be measured, so they are calculated, with results as shown in Table XI.

Table XI. Equilibrium composition of hydrogen plasma (basis of 1 mole of cold H_2)

		Temperature (K)		
p(atm)	Component	5000	7000	9000
10	H_2	0.28644	1.888×10^{-2}	3.216×10^{-3}
	H	1.4271	1.9619	1.9875
	H^+	1.716×10^{-6}	3.206×10^{-4}	6.023×10^{-3}
1	H_2	4.497×10^{-2}	1.926×10^{-3}	3.164×10^{-4}
	H	1.9101	1.9952	1.9817
	H^+	6.654×10^{-6}	9.929×10^{-4}	1.771×10^{-2}
0.1	H_2	4.785×10^{-3}	1.933×10^{-4}	3.009×10^{-5}
	H	1.9904	1.9965	1.9464
	H^+	2.156×10^{-5}	3.078×10^{-3}	5.353×10^{-2}
0.01	H_2	4.809×10^{-4}	1.588×10^{-5}	2.541×10^{-6}
	H	1.9990	1.9904	1.8359
	H^+	6.824×10^{-5}	9.604×10^{-3}	0.16405

Source: Reference 8.

When a spacecraft reenters the atmosphere at high velocities, shocks develop that heat the surrounding air to such high temperatures that ionization occurs, resulting in a brief blackout of radio communication. To understand these phenomena, one needs the properties of very hot air. Typical results are shown in Table XII.

Table XII. Equilibrium composition and properties of air at 12,000 K and 9.0615 atm

Component	Moles present in plasma
N	1.4848
N^+	0.06586
O	0.39627
O^+	0.023048
Ar	0.8960×10^{-2}
Ar^+	0.3720×10^{-3}
N_2	0.49975×10^{-2}
N_2^+	0.20138×10^{-3}
NO	0.31663×10^{-3}
NO^+	0.26067×10^{-3}
e	0.089746
Enthalpy (kcal)	173.97
Entropy (cal/K)	109.59
Gibbs free energy (kcal)	−1,141.1

Source: Reference 8.

The table gives the number of moles of each of 11 species present at one T and p beginning with one mole of dry air as a ternary mixture at STP. The equilibrium composition of the plasma was determined by a minimization of the free energy using the technique of Lagrangian undetermined multipliers. All properties were determined using the techniques of statistical thermodynamics. All other possible species were calculated to be 0.1 times as abundant as the least abundant species in the table. The calorimetric properties of most of these species are not experimentally accessible. We believe the calculated properties to be accurate, however, for with other molecules and in regions where calculated and experimental properties may be compared, the agreement is good, as we have seen.

Frequencies and moments from quantum mechanics (9)

In all of our previous arguments, we have taken the molecular parameters of moments of inertia and vibrational frequencies as experimental quantities from which the formalism of statistical mechanics allowed us to deduce all of the thermodynamic properties. We have marvelled on several occasions at how *optical* measurements could be used to rigorously deduce *calorimetric* or *thermal* quantities. Statistical mechanics provides the bridge between these molecular properties on the one hand and the macroscopic behavior of matter on the other hand. And the essential component of this bridge is the partition function. A legitimate question now is, "Is it possible to calculate from first principles—from quantum mechanics—the molecular structure and vibrational frequencies?" This is, of course, an ultimate goal of theoretical chemistry: Calculate all of the macroscopic properties of matter from first principles without having to measure anything experimentally. The calculation must also be cost-effective; a precise heat of formation obtained at a greater cost than by an experimental measurement has accomplished little. Approximate quantum mechanical insights of sufficient power and accuracy do exist to enable calculations of ΔH_f^0, molecular structure, and vibrational frequencies. Hence we are able to immediately calculate all of the thermodynamic properties of the ideal gas at all temperatures and pressures with no required input data at all.

Simple Huckel techniques reveal the essence of the problem. More serious attacks center upon either ab initio techniques or semiempirical techniques. The latter have been highly developed, and it is possible to calculate observable properties to within useful accuracies. Heat capacities good to within a joule, absolute entropies to within about a J/mol K, and heats of formation to within a few kJ or so are common. The required input information is only the chemical formula of the molecule. The computer codes are available (10), and they can be used more or less as "black boxes," for it is not at all necessary to understand the intricacies of the quantum mechanical arguments in order to use the techniques. (By analogy, one need not understand electronic pulsing circuitry in order to use a mass spectrometer.)

An examination of the ideas and approximations of semiempirical quantum mechanical arguments would take us too far into theoretical chemistry. It suffices merely to note that in exactly the same way that electrons about an atom are envisioned as occupying atomic orbitals (AOs), so we assume that the valence electrons of molecules occupy molecular orbitals (MOs). These MOs are described mathematically as a linear combination of the AOs of each atom, and these may be reasonably approximated. We take these outer electrons to move in a fixed field formed by the nuclei and the inner core of electrons. These assumptions together form the so-called linear combination of atomic orbitals, molecular orbital, self-consistent field approach (or LCAO-MO-SCF) to molecular quantum mechanics.

In quantum mechanical calculations we must deal with many complex integrals that may be difficult or impossible to evaluate accurately, so we assign values to integrals or parameterize the formalism to force a fit to the known heats of atomization and geometries of a set of standard molecules that has been chosen to represent as wide a variety of types of structures and bonding as possible. To illustrate the idea in a greatly oversimplified setting, if we parameterize on the experimental facts of CH_4 and C_2H_6 to characterize the C—H and C—C bonds respectively, we can then use the now calibrated theory to calculate the properties of all straight chain alkanes.

There are a number of common criticisms of semiempirical methods. By their nature, such methods are curve-fitting and should not be relied upon outside the range of compounds used to determine the parameters. Also by their nature, such methods can be reparameterized indefinitely, and several versions of each technique have appeared. Some early versions are now out of date. It appears that current parameterization is about as good as it can be relative to the basic quantum mechanical assumptions that are involved in each method. We expect little further improvement due to reparameterization until better theory is evolved. Certainly it is necessary to use discretion in interpreting results; one must reason analogously; one must use one's chemical intuition; and one must be skeptical.

Quantum mechanical calculations permit the calculation of the energy of a molecule as a function of its geometry; that geometry which minimizes the energy of a molecule is usually close to the actual geometry of the molecule. We must define a zero of energy someplace, and this standard state is almost always that state of aggregation of the pure element at 25 °C and 101.325 kPa (1 atm). But the quantum mechanical arguments give us the energy of the molecule relative to its separated atoms. Thus, to write the usual thermochemical heat of formation, we need also to know the heats of formation of atoms as ideal gases at 25 °C and 101.325 kPa from the elements in their standard states. This involves well-known quantities like the bond energy of H_2 and the heat of sublimation of graphite. So the quantum mechanically calculated total energy of the molecule can be converted into its heat of formation, ΔH_f, which is the same quantity that one would measure calorimetrically.

The calculated geometry yields the moments of inertia, while calculating what is called the force constant matrix yields the vibrational frequencies.

Some comparisons with experiment

The usefulness of any theory is determined by its comparison with experiment. Figure 3 shows a plot of calculated versus experimental ΔH_f^0 (at 25 °C) for 193 molecules. The agreement is quite good, with most of the compounds lying within ±5 kcal/mol (±20 kJ/mol) of the line of unit slope. The types of compounds included in Figure 3 are ethane, ethylene, acetylene, propene, isomeric butanes, butadiene, cyclopropanes, cyclopropenes, cyclobutanes, cyclopentanes, cyclobutenes, cyclopentenes, cyclohexanes, substituted benzene derivatives, amines, aldehydes, ketones, nitrates, phenols, fluorinated hydrocarbons, silanes, phosphanes, sulfur-containing compounds, etc., and also free radicals and carbonium ions. In short, the heats of formation of a wide variety of molecules have been well calculated. Some more detailed comparisons of experiment and theory appear in Table XIII.

Figure 3. Plot of calculated against observed heats of formation of 193 compounds derived from H, C, N, O, F, Si, P, S, and Cl. The line is the theoretical line of unit slope, not one drawn through the points. *Adapted with permission from Ref. 11.*

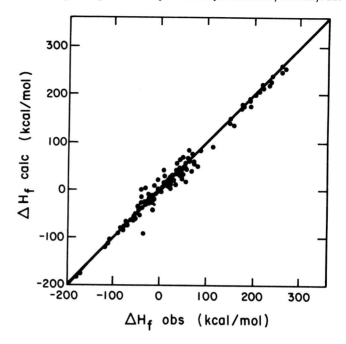

In addition, the calculated geometries were found to be in good agreement with experiment. Experimental problems are such that discrepancies with the theory of less than 0.001 nm in bond length and 1 degree in bond angle are not too meaningful. Mean errors in bond lengths were 0.0014 nm and 2.8 degrees in bond angle in calculations using a semiempirical code called MNDO. Examples of agreement for several species appear in Table XIV. Moments of inertia are, of course, immediately calculated from the geometry, and Figure 4 shows calculated versus experimental moments of inertia for many of the same molecules that were

Figure 4. Plot of calculated vs. observed moments of inertia for 27 compounds. *Adapted with permission from Reference 12.*

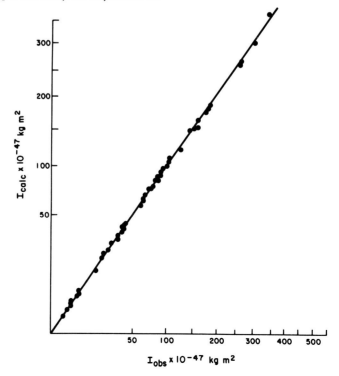

Table XIII. Heat of formation calculated using MNDO vs. the experimental values

Molecule	ΔH_f^0 kcal/mol			Molecule	ΔH_f^0, kcal/mol		
	Calc	Exp	Error		Calc	Exp	Error
(iPr-N=N-iPr)	2.2	8.6	-6.4	C_2H_5CHO	-48.1	-45.5	-2.6
				$(CH_3)_2CO$	-49.5	-51.9	+2.4
				CH_2CO	-7.0	-11.4	+4.4
CH_2N_2	67.1	71	-3.9	$(CHO)_2$	-61.6	-50.7	-10.9
				$(CH_3CO)_2$	-78.9	-78.2	-0.7
(diazirine)	72.5	79	-6.5	$CH_3COCH_2COCH_3$	-83.3	-90.5	+7.2
				(O=C6H4=O)	-33.1	-29.3	-3.8
HN_3	73.0	70.3	+2.7				
O_2	12.2	22.0	-9.8	(Ph-CHO)	-9.8	-8.8	-1.0
O_3	48.5	34.2	14.3				
H_2O	-60.9	-57.8	-3.1	HCOOH	-92.7	-90.6	-2.1
CH_3OH	-57.4	-48.1	-9.3	CH_3COOH	-101.2	-103.3	+2.1
C_2H_5OH	-63.0	-56.2	-6.8	C_2H_5COOH	-105.7	-108.4	+2.7
n-C_3H_7OH	-67.7	-61.2	-6.5	$(COOH)_2$	-174.3	-175.0	+0.7
i-C_3H_7OH	-65.4	-65.1	-0.3				
t-C_4H_9OH	-64.3	-74.7	+10.4	(Ph-COOH)	-65.9	-70.1	+4.2
CH_3OCH_3	-51.2	-44.0	-7.2				
$C_2H_5OC_2H_5$	-62.0	-60.3	-1.7	$HCOOCH_3$	-85.6	-83.6	-2.0
(cyclopropane-O)	-15.5	-12.6	-2.9	CH_3COOCH_3	-93.6	-97.9	+4.3
				$CH_3CO-O-COCH_3$	-132.8	-137.1	+4.3
(THF)	-8.7	-8.3	-0.4				
				(maleic anhydride)	-88.7	-95.2	+6.5
(Ph-OH)	-24.7	-23.0	-1.7				
(Ph-OCH3)	-18.0	-17.3	-0.7	$HCONH_2$	-39.8	-44.5	+4.7
HO—OH	-38.2	-32.5	-5.7	$HCON(CH_3)_2$	-36.9	-45.8	+8.9
CH_3O-OCH_3	-28.3	-30.1	+1.8	N_2O	30.9	19.6	+11.3
$C_2H_5O-OC_2H_5$	-39.1	-46.1	+7.0	(HONO)	-40.6	-18.8	-21.8
CO	-6.2	-26.4	+20.2				
CO_2	-75.4	-94.1	+18.7	$HONO_2$	-17.5	-32.1	+14.6
C_3O_2	-24.1	-22.4	-1.7	CH_3ONO	-36.6	-15.8	-20.8
CH_2O	-33.0	-26.0	-7.0	CH_3NO_2	3.3	-17.9	+21.2
CH_3CHO	-42.4	-39.7	-2.7				

Source: Reference 7.

Table XIV. Optimized structures for several molecules as calculated by MNDO. Experimental values appear in parentheses.

Molecule	Bond length (Å)	Bond angle
$C(CH_3)_4$	CC 1.554(1.539) CH 1.109(1.120)	HCC 111.7(110.0)
(cyclobutane)	CC 1.549(1.548) CH 1.105(1.133)	HCH 107.6(108.1) $C^1C^2C^4C^3$ 180
(benzene)	CC 1.407(1.397) CH 1.090(1.084)	
(ketene-like)	CO 1.417(1.435) CC 1.513(1.470)	HCH 111.7(116.3) C—CH_2 159.2(158.1)
(naphthalene)	C^1C^2 1.382(1.364) C^2C^3 1.429(1.415) C^1C^9 1.439(1.421) C^9C^{10} 1.435(1.418)	

Source: Reference 7.

depicted in Figure 3. The logarithmic scale depicts the great range of values that have been successfully calculated. In view of this variety of molecules, the results are remarkably good.

The computational scheme is to input a trial geometry of the molecule of interest. Which atom is connected to which and its approximate distance and angles from the other atoms are all that is required. Cartesian coordinates of each atom are not necessary. The semiempirical quantum mechanical code called MNDO will optimize the geometry for minimum total energy, calculate the ΔH_f^0, and calculate the coordinates of each atom (*10*). A second code, called GEOMO/RV, takes the MNDO geometry as input and calculates vibrational frequencies and moments of inertia. A third code, a statistical thermodynamics package, takes the output of GEOMO/RV and calculates the thermodynamic properties.

Figure 5 shows a plot of nearly 500 calculated versus experimental vibrational frequencies for 34 molecules, with an average error of about ±10%. These include water, hydrogen sulfide, ammonia, carbon dioxide, carbon disulfide, hydrogen cyanide, formaldehyde, methanol, ethylene, methylamine, dimethylether, acetone, maleic anhydride, furan, pyrrole, thiophene, benzene, and a variety of heterocyclic compounds.

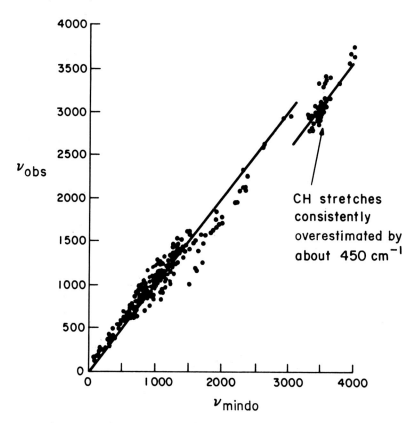

Figure 5. Calculated vs. observed vibration frequencies for molecules. *Adapted with permission from Reference 13.*

A wide variety of additional properties of molecules have been investigated—including dipole moments, first ionization potentials, polarizabilities, nuclear quadrupole coupling constants, ESCA chemical shifts, and the electronic band structure of polymers. In addition, the mechanisms predicted for several hundred reactions have also been consistent with available experimental data.

As a natural extension and the one most important for our purposes here, the heat capacities and absolute entropies were also calculated using the above predictions of moments and frequencies. A comparison of results is presented in Table XV, where the agreement is seen to be excellent, being within 1 cal/mol K (4.2 J/mol K) at 298 K in almost every case.

Finally, the free energies of reaction ΔG^0 have been calculated for several industrially important reactions at various temperatures, and comparisons with experiment are presented in Table XVI. The entropies of the various compounds were calculated directly at 300, 900, and 1500 K. The enthalpies of reaction were calculated from experimentally determined ΔH_f^0 data but corrected to the appropriate temperature using $\Delta H_T^0 - \Delta H_{298}^0 = \sum (H^0 - H_0^0)_T - \sum (H^0 - H_0^0)_{298}$ and a semiempirical quantum mechanical code called MINDO/3 to calculate values of $(H^0 - H_0^0)$ (9). The agreement with experiment is quite remarkable. Recall, however, that an error of a few kcal in ΔG^0 may cause order of magnitude errors in the equilibrium constant.

Table XV. Calculated and observed entropies and heat capacities at 298 K.

Compound	S^0, cal mol^{-1} deg^{-1}			C_p^0, cal mol^{-1} deg^{-1}		
	Obs	Calc	Error	Obs	Calc	Error
H_2O	45.1	45.0	0.1	8.0	8.0	0.0
H_2S	49.1	49.2	−0.1	8.2	8.5	−0.3
NH_3	46.0	46.0	0.0	8.5	8.4	0.1
CO_2	51.1	51.4	−0.3	8.9	9.2	−0.3
CS_2	56.8	57.4	−0.6	10.9	11.3	−0.4
HCN	47.9	48.1	−0.2	7.8	8.4	−0.6
H_2CO	55.3	55.0	0.3	10.1	10.5	−0.4
CH_2N_2	58.0	58.7	−0.7	12.5	13.0	−0.5
CH_3Cl	55.9	56.1	−0.2	9.7	10.4	−0.7
CH_4	44.5	44.6	−0.1	8.5	8.7	−0.2
HC≡CH	48.0	48.3	−0.3	10.5	10.6	−0.1
CH_2=C=O	57.8	58.3	−0.5	12.4	12.9	−0.5
CH_3CN	58.1	58.1	0.0	12.5	12.7	−0.2
CH_2=CHCl	63.1	63.9	−0.8	12.8	13.8	−1.0
CH_2=CH_2	52.4	52.6	−0.2	10.2	11.0	−0.8
oxirane (O)	58.0	58.0	0.0	11.4	11.6	−0.2
thiirane (S)	61.9	61.3	0.6	13.3	13.7	−0.4
aziridine (NH)	59.8	60.0	−0.2	12.2	12.9	−0.7
H_2C=C=CH_2	58.2	58.6	−0.4	14.1	14.7	−0.6
CH_3C≡CH	59.3	59.3	0.0	14.5	14.5	0.0
cyclopropane	56.6	56.9	−0.3	13.1	13.7	−0.6
maleic anhydride	72.6	74.5	−1.9	20.7	21.4	−0.7
furan	63.8	64.8	−1.0	15.6	17.0	−1.4
thiophene	67.9	68.3	−0.8	17.4	19.3	−1.9
pyrrole	65.8	66.4	−0.6	16.8	19.7	−1.9
isoxazole	64.8	65.0	0.0	14.5	14.8	−0.3
benzene	64.3	65.2	−0.9	19.6	20.9	−1.3

Source: Reference 12.

Table XVI. Free energies of reaction for several industrially important reactions calculated from molecular orbital theory but using experimental heats of formation at 298 K

Reaction	ΔH^0(reaction)$_{298}$ (from experimental data)	Gibbs Free Energy of reaction for ideal gas at 1 atm (ΔG^0 kcal/mole)					
		300 K		900 K		1500 K	
		Obs	Calc	Obs	Calc	Obs	Calc
$N_2 + 3H_2 \rightarrow 2NH_3$	−21.94	−7.74	−7.82	24.24	24.15	58.36	57.55
$2CH_4 \rightarrow CH$≡$CH + 3H_2$	88.80	68.01	68.04	31.65	31.91	−7.43	−7.01
CH_2=CH_2 + 0.5 O_2 → H_2C—CH_2 (O)	−25.12	−19.45	−19.72	−7.70	−8.07	3.88	5.00
$CH_4 + 2H_2O \rightarrow 4H_2 + CO_2$	39.45	27.07	27.03	−2.02	−1.80	−34.25	−33.72
$CH_4 + NH_3 + 1.5\ O_2 \rightarrow HCN + 3H_2O$	−112.33	−118.02	−119.02	−129.48	−133.81	−139.83	−143.43
$CH_4 + Cl_2 \rightarrow CH_3Cl + HCl$	−24.82	−25.64	−25.60	−27.48	−27.52	−29.46	−29.78

Source: Reference 12.

An Example:

A number of known compounds of N, O, and F are powerful oxidizers. Many more similar molecules may be postulated. For a potential oxidizer for a rocket engine, a high positive heat of formation is desired, for then the reaction would yield the hottest possible flame gases. But a high positive ΔH_f^0 also implies lack of stability. Can quantum arguments suggest which species might be more profitable targets of an expensive program in synthetic chemistry?

Theoretically predicted heats of formation of 7 known N-O-F compounds as well as 11 postulated compounds appear in Table XVII. Calculations employed the MINDO/3 semiempirical scheme (14). No data on any compound in this table were used to calculate any of the values of ΔH_f^0. When this study was performed, only ΔH_f^0 could be calculated using available codes. As a qualitative index of the relative stability of these compounds, let us estimate the heats of reaction for the dissociation of each compound into reasonable fragments. If this heat of reaction is positive, we might ascribe a modest possible stability to the postulated compound. On this basis, FONO, FO_2NO, $F_2N(OF)$, and $F_2N(O_2F)$ were the most likely candidates for an experimental synthetic effort. Interestingly, FONO was experimentally observed in another laboratory about the same time that these predictions were being developed. Even so, it is questionable whether any of these compounds would be interesting rocket fuel oxidizers because of their evident low endothermicity. But this too is a practical conclusion from quantum mechanics.

Table XVII. Theoretical and experimental heats of formation of several known and postulated N-O-F compounds. Experimental values on all of the known compounds are in parentheses.

Compound	ΔH_f^0 kcal/mol	Compound	ΔH_f^0 kcal/mol
FNO_2	−31.48 (−25.8 to −33.8)	F_2NONF_2	21.69
$FONO_2$	1.5 (2.5)	$F_2NO_2NF_2$	37.12
ONF_3	−33.31 (−34.1)	ONNFNFNO	78.8
$ONNF_2$	18.21 (18.9)	$F_2N(OF)$	−20.48
O_2NNF_2	−14.67 (0)	$FN(OF)_2$	−9.87
FNNF	19.45 (19.4)	$N(OF)_3$	6.56
F_2NNF_2	−2.81 (−5.0)	$F_2N(O_2F)$	7.88
FONO	7.59	$FN(O_2F)_2$	74.02
FO_2NO	−31.1	$N(O_2F)_3$	140.1

Source: Reference 14.

Summary

Quantum mechanics and statistical mechanics can provide answers to practical problems. The values calculated for properties like ΔH_f^0, S^0, C_p^0, and ΔG^0 are obviously important. We now have insight, for example, into why the equilibrium constant for one reaction is large while that for another is small. These same techniques may be readily applied to solids. These techniques may even be generalized to enable insights into the properties of real gases and liquids, but this requires rather more tedious conceptual understanding. Even more to the point, these results have shown how very esoteric concepts can be taken out of the "in principle" category and placed into the "in practice" category. In a totally different field of science, the very title of a book, *Quantum Pharmacology* (15), is further evidence of the real-world utility of quantum mechanics. Surely such realizations will remove the mysticism from the study of these otherwise esoteric subject areas.

References
(1) Herzberg, G. *Atomic Spectra and Atomic Structure*; Dover: New York, 1944.
(2) Moelwyn-Hughes, E. A. *Physical Chemistry*; Pergamon: New York, 1957.
(3) Landolt-Bornstein. *Kalorische Zustandsgrossen*; Springer-Verlag: Berlin, 1961.
(4) Chao, J.; Zwolinski, B. J. *J. Phys. Chem. Ref. Data* **1975**, *4*, 251.
(5) Fateley, W. A.; Miller, F. A. *Spectrochim. Acta* **1962**, *18*, 977.
(6) Li, J.C.M. *J. Am. Chem. Soc.* **1956**, *78*, 1081.
(7) Dewar, M.J.S.; Thiel, W. *J. Am. Chem. Soc.* **1977**, *99*, 4907.
(8) McGee, H. A., Jr.; Heller, G. *Progress in Astronautics and Aeronautics*, Vol. 9; Academic: New York, 1963.
(9) Shah, K. N. MS Thesis, Virginia Polytechnic Institute and State University, August 1983.
(10) Request the MINDO, MNDO, or similar semiempirical codes from the Quantum Chemistry Program Exchange, Indiana University, Bloomington.
(11) Bingham, R. C., et al. *J. Am. Chem. Soc.* **1975**, *97*, 1285.
(12) Dewar, M.J.S.; Ford, G. P. *J. Am. Chem. Soc.* **1977**, *99*, 7822.
(13) Dewar, M.J.S.; Ford, G. P. *J. Am. Chem. Soc.* **1977**, *99*, 1685.
(14) Ganguli, P. S.; McGee, H. A., Jr. *Inorg. Chem.* **1972**, *11*, 3071.
(15) Richards, W. G. *Quantum Pharmacology*; Butterworths: Boston, 1977.

Chapter 6   MOLECULAR THERMODYNAMICS OF FLUID MIXTURES

J. M. PRAUSNITZ
Chemical Engineering Department
University of California
Berkeley, Calif. 94720

We live in a world of mixtures. The air we breathe, the blood in our veins, the petroleum we extract from the earth, the water in the oceans, the juice we press out of oranges—wherever we turn, we encounter mixtures, mostly fluid mixtures. It is important to understand the properties of such mixtures because they are required in a variety of fundamental sciences, including physiology, meteorology, and oceanography, as well as in engineering sciences based on physical chemistry. This chapter is concerned with some equilibrium properties of a few typical fluid mixtures.

About 100 years ago, in the early days of physical chemistry, much attention was given to the equilibrium properties of fluids, pure and mixed. This early attention was stimulated, first, by J. W. Gibbs at Yale, who showed that thermodynamics (traditionally a science for heat engines) could be extended to provide a rigorous framework for calculating phase equilibria; and second, by the work of J. D. van der Waals at Amsterdam, who proposed a remarkably simple molecular theory which (approximately) relates macroscopic properties of pure fluids and their mixtures to molecular sizes and intermolecular forces. Although the work of Gibbs was essentially complete 100 years ago, that of van der Waals has been subjected to extensive refinement, which continues today. Van der Waals's essential ideas have remained useful for over a century; they are the basis of much that is presented here.

Molecular thermodynamics

Classical thermodynamics is revered, honored, and admired for its intellectual beauty. In practice, however, it is inadequate; because classical thermodynamics is independent of any theory of matter, it can interrelate but not quantitatively predict thermodynamic properties without extensive experimental information.

Classical thermodynamics's independence of a molecular theory of matter provides its glory, but this independence is also responsible for its weakness. In his later work, Gibbs tried to overcome this weakness through application of statistical physics as developed by Boltzmann, leading to statistical mechanics, which is a rigorous attempt to describe bulk (macroscopic) properties of matter in terms of (microscopic) intermolecular forces. Although statistical mechanics (the basis of statistical thermodynamics) has made impressive progress since the early efforts of Gibbs and Boltzmann, it has not yet reached a stage where, by itself, it can yield practical results for calculating the properties of real mixtures.

Statistical thermodynamics alone cannot provide most desired mixture properties. It does, however, suggest techniques for constructing sensible physical models which, when coupled with appropriate concepts from physical chemistry, can be used to correlate limited experimental data. Such correlations very much reduce experimental effort by providing good estimates of bulk properties through physically meaningful interpolation and extrapolation.

Molecular thermodynamics is a combination of classical thermodynamics, statistical thermodynamics, molecular physics, and physical chemistry. It is a compromise and a synthesis; its methodology is indicated by the following steps:

1. Use statistical thermodynamics whenever possible, at least as a point of departure.
2. Apply appropriate concepts from molecular physics and physical chemistry.

3. Construct physically grounded models for expressing abstract thermodynamic functions in terms of real, measurable properties.
4. Obtain model parameters from only a few, but representative, experiments.

Generalized van der Waals partition function

As any text on statistical thermodynamics (1, 2) shows, every thermodynamic property can be calculated from partition function Q, which is a function of total volume V containing N_1 molecules of component 1, N_2 molecules of component 2, etc., at temperature T. For phase equilibrium calculations, the important properties are pressure P and, for each component i, chemical potential μ_i. These are related to Q through

$$P = kT\left(\frac{\partial \ln Q}{\partial V}\right)_{T,\, \text{all } N} \tag{1}$$

$$\mu_i = -kT\left(\frac{\partial \ln Q}{\partial N_i}\right)_{T, V, N_j} \tag{2}$$

where k is Boltzmann's constant.

Except for ideal gases or very simple solids, there is no rigorous and practical method for expressing Q as a function of V, T, and N_i, N_j.... However, an approximate method can be developed from the ideas van der Waals expressed in his doctoral thesis of 1873.

Essentially, van der Waals generalized for a real fluid the rigorous expression for an ideal gas. For a pure, ideal gas

$$Q = \frac{1}{N!}\left[\frac{V}{\Lambda^3}\right]^N [q_{r,v,e}]^N \tag{3}$$

where the de Broglie wavelength Λ is given by

$$\Lambda = h(2\pi mkT)^{-1/2} \tag{3a}$$

with m = molecular mass and h = Planck's constant.

Here $q_{r,v,e}$ stands for the partition function of the isolated molecule due to rotational, vibrational, and electronic degrees of freedom. The term V/Λ^3 gives the contribution to Q arising from three translational degrees of freedom.

For a real, pure fluid the van der Waals partition function becomes

$$Q = \frac{1}{N!}\left[\frac{V_f}{\Lambda^3}\right]^N \left[\exp\frac{-\Phi}{2kT}\right]^N [q_{r,v,e}]^N \tag{4}$$

Here V_f is the total free volume and $\Phi/2$ is the potential field experienced by one molecule due to attractive forces from all the others. Equation 4 is not rigorous, but for many cases it provides an excellent approximation.

In an ideal gas, molecules are points that do not influence each other. However, in a real fluid, molecules have a finite size, leading to repulsive forces at close molecule-molecule separations; further, they have electronic configurations, leading to forces of attraction at intermediate separations. In the generalized van der Waals model, repulsive forces are taken into account through the concept of free volume V_f, and attractive forces are taken into account through the concept of potential field Φ.

In addition to translational motions, molecules also have rotational and vibrational degrees of freedom that contribute to the partition function. (Electronic degrees of freedom are important only at high temperatures and therefore are not of concern here.) For simple, small (argon-like) molecules, it is reasonable to assume that these contributions depend only on temperature. In that event, they do not contribute to the equation of state (Equation 1); further, in phase equilibrium calculations, where all phases are at the same temperature, these contributions cancel out.

Free volume

In van der Waals's theory of fluids, volume V is replaced by free volume V_f to allow for the effect of repulsive forces at small intermolecular distances. Since real (as opposed to ideal) molecules have a finite size, two molecules that "touch" strongly repel each other; if the molecules are "hard" (i.e., incompressible), then, upon touching, they repel with an infinitely large force. (If the molecules are "soft," the repulsive force is large but not infinite.)

Figure 1 gives a two-dimensional view of the free-volume concept. The free volume is that part of the total volume that is accessible; because of the finite size of all molecules, a particular molecule, wandering about, does not have access to every part of the total volume since some of that volume (the excluded volume) is occupied by other molecules.

Figure 1. Free volume for an assembly of spheres with and without overlap.

FREE VOLUME

FREE VOLUME = TOTAL VOLUME − EXCLUDED VOLUME

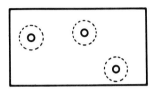

DILUTE REGION (NO OVERLAP)
$$V_f = V - \frac{N}{N_A} b$$

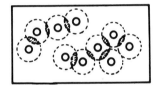

DENSE REGION (EXTENSIVE OVERLAP)
$$V_f > V - \frac{N}{N_A} b$$

In Figure 1, each molecule is represented by a billiard ball, or hard sphere. The finite size of a molecule is indicated by the dark line. Let one molecule approach another until the two surfaces are in contact; this is the closest possible molecule-molecule distance. The center of one of the molecules cannot penetrate the volume bounded by the dashed line around the other. That volume is not available; it is excluded. When the number of molecules is small (low density), this excluded volume (per mole) is given by van der Waals parameter b, which is 4 times the actual volume of Avogadro's number of finite-sized molecules [$b = (2/3)\pi N_A \sigma^3$]. Here σ is the molecular diameter.

At high densities, it is much more difficult to calculate the excluded volume because of overlap, as shown by the shaded areas in the lower part of Figure 1. The excluded volume (per mole) is less than b because, in the calculation of b, the overlapping regions are now erroneously counted more than once. Van der Waals and his co-workers were well aware of overlap, but no satisfactory solution to this difficult mathematical problem was achieved until the middle 1960s. There is no need to go into details here; it is sufficient to say that for hard-sphere molecules we now have a reliable expression for the free volume for all fluid densities:

$$V_f = V \exp\left[\frac{(\tau/\tilde{v})(3\tau/\tilde{v} - 4)}{(1 - \tau/\tilde{v})^2}\right] \tag{5}$$

Equation 5 is a form of the Carnahan-Starling equation. Here τ is a numerical constant $= \pi\sqrt{2}/6 = 0.7405$, and $\tilde{v}$ is a reduced volume given by $\tilde{v} = V/Nv^*$ where $v^* = \sigma^3/\sqrt{2}$.

Potential energy

Figure 2 gives a two-dimensional view of the potential-field concept. Consider the molecule at the center; we want to calculate the attractive-force field experienced by that molecule. The potential energy for any pair of molecules is designated by $\Gamma(r)$, where r is the center-to-center distance of separation. To sum up the contributions of all molecules interacting with that at the center, we perform an integration, as shown. Imagine a sphere of radius r; the area of that sphere is $4\pi r^2$. We now build a thin shell on that area; the thickness of the shell is dr and its volume is $4\pi r^2 dr$. We calculate the number of molecular centers in the shell by multiplying $4\pi r^2 dr$ by the overall density N/V; thereby we obtain the total potential experienced by the central molecule.

Figure 2. Potential energy is the sum of pair potentials.

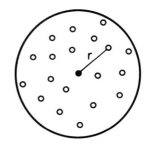

POTENTIAL ENERGY Φ

POT. ENERGY BETWEEN TWO MOLECULES IS $\Gamma(r)$.
POT. ENERGY BETWEEN CENTRAL MOLECULE AND ALL OTHERS IS

$$\Phi = \frac{N}{V} \int_0^\infty \Gamma(r) g(r) 4\pi r^2 \, dr$$

$g(r)$ = RADIAL DISTRIBUTION FUNCTION, TELLS US WHERE THE MOLECULES ARE

Unfortunately, there is a crude calculation that erroneously assumes that the local density (the density of molecules at position r given that there is a molecule at the center) is everywhere equal to the overall (or average) density. The structure of a fluid is not completely random; because of the finite size of the molecules, there is short-range order. When r is less than the molecular diameter, the local density is zero; when r is only slightly larger than the molecular diameter, the local density is larger than the average density. Therefore, it is necessary to put into the integral some structural information given by the radial distribution function $g(r)$, which is the ratio of local density to average density. The notation $g(r)$ is deceptive because it erroneously implies that g is a function only of r; in fact, it is a strong function of density N/V and, to a lesser extent, of temperature.

The last 25 years have produced significant progress in our understanding of the radial distribution function, especially because of the computer simulation (molecular dynamics) studies of Alder et al. (*3*). Again, there is no need to go into details; it is sufficient to say that we now have some useful theoretical methods for calculating Φ as a function of density and temperature.

Table 1 summarizes some expressions for V_f and Φ that lead to well-known semiempirical equations of state. In addition, Table 1 gives an expression for Φ that is based on computer simulation studies for spherical molecules with a square-well potential. A square-well potential is described by

$$\Gamma(r) = \begin{cases} \infty & \text{for } r \leq \sigma \\ -\epsilon & \text{for } \sigma < r \leq R\sigma \\ 0 & \text{for } r > R\sigma, \end{cases}$$

where $(R - 1)\sigma$ is the well width.

Table I. Some expressions for free volume V_f and for potential Φ leading to equation of state (EOS)

Author	V_f	$-\Phi$	EOS
van der Waals	$V - \dfrac{Nb}{N_A}$	$\dfrac{2aN}{VN_A^2}$	$z = \dfrac{v}{v-b} - \dfrac{a}{RTv}$
Redlich-Kwong	$V - \dfrac{Nb}{N_A}$	$\left(\dfrac{T_c}{T}\right)^{1/2} \dfrac{N_A V}{Nb} \times \ln\left(1 + \dfrac{Nb}{N_A V}\right)$	$z = \dfrac{v}{v-b} - \dfrac{a}{RT^{3/2}(v+b)}$
Eyring	$8\left[V^{1/3} - \left(\dfrac{3Nb}{2\pi N_A}\right)^{1/3}\right]^3$	$\dfrac{2aN}{VN_A^2}$	$z = \dfrac{v}{v - \left(\dfrac{3b}{2\pi}\right)^{1/3} v^{2/3}} - \dfrac{a}{RTv}$
Alder	Equation 5	$2\epsilon \sum_{n=1}^{4} \sum_{m=1}^{M} \left(\dfrac{A_{nm}}{\tilde{v}^m}\right) \times \left(\dfrac{1}{\tilde{T}^{n-1}}\right)$	$z = z$ (Carnahan-Starling) $- \dfrac{\text{``}a\text{''}}{RTv}$ $z$ (Carnahan-Starling) $= 1 + \dfrac{4(\tau/\tilde{v}) - 2(\tau/\tilde{v})^2}{(1 - \tau/\tilde{v})^3}$ $\text{``}a\text{''} = -N_A \epsilon v^* \sum_{n=1}^{4} \sum_{m=1}^{M} \times \left(\dfrac{m A_{nm}}{\tilde{v}^{m-1}}\right)\left(\dfrac{1}{\tilde{T}^{n-1}}\right)$

$b = (2/3)\pi N_A \sigma^3$ where N_A = Avogadro's number; $z = Pv/RT$; v = volume per mole = VN_A/N; a is a characteristic positive constant; T_c is the critical temperature; $\tilde{T} = kT/\epsilon$, where ϵ is a characteristic potential-energy parameter. Generalized coefficients A_{nm} and M are given in Reference 3. "a" is an effective parameter depending on $\tilde{v}$ and $\tilde{T}$ where $\tilde{v} = v/N_A v^*$ with $v^* = \sigma^3/\sqrt{2}$; $\tau = 0.7405$.

Extension to mixtures

There is no practical rigorous way to extend Equation 4 to mixtures, but there are several phenomenologically reasonable ways to do so. As a first step, we can write for the mixture

$$Q = \prod_i \frac{1}{N_i!} \left[\frac{\langle V_f \rangle^N}{3N_i \prod_i \Lambda}\right] \left[\exp \frac{-\langle \Phi \rangle}{2kT}\right]^N \left[\prod_i q_{(r,v,e)_i}^{N_i}\right] \qquad (6)$$

where $N = \sum_i N_i$ and the brackets $\langle \; \rangle$ stand for a composition average.

Numerous suggestions have been made for calculating $\langle V_f \rangle$ and $\langle \Phi \rangle$; these suggestions (known as mixing rules) are almost always in part empirical although they can often be supported by theory, at least for mixtures of spherical molecules. The simplest nontrivial mixing rules are

$$\langle V_f \rangle = V - \sum_i \frac{N_i}{N_A} b_i \qquad (7)$$

$$\langle \Phi \rangle = \sum_i \sum_j x_i x_j \Phi_{ij} \qquad (8)$$

where x is the mole fraction. (Equation 7 neglects overlap.)

Equation 7 is simple because it requires only pure-component data: For each component i, we require size parameter b_i. However, for a binary mixture, Equation 8 requires potential Γ_{12} and radial distribution function g_{12} in addition to pure-component data. Since these are not easily determined, the simplest nontrivial approximation is to assume that

$$\Phi_{ij} = (\Phi_{ii}\Phi_{jj})^{1/2}(l - k_{ij}) \qquad (9)$$

where empirical binary constant k_{ij} may depend on temperature but not on density or composition. For simple mixtures, k_{ij} is often zero. For many typical mixtures of nonpolar nonelectrolytes, k_{ij} is positive and of order 0.01–0.10.

To illustrate the use of generalized van der Waals theory, we consider vapor-liquid equilibria for the mixture methane-propane. To describe this mixture, we use the Carnahan-Starling equation for free volume and an expression for Φ based on computer simulations by Alder (see Table 1). In this case, the Carnahan-Starling equation for pure fluids was extended to mixtures by using the so-called one-fluid assumption: The free volume of the mixture is assumed to be that of a hypothetical pure fluid whose molecular size is some composition average of the molecular sizes for the mixture's components. For this mixture, molecular size parameter b was calculated from

$$b \text{ (mixture)} = \sum_i x_i b_i \qquad (10)$$

(The one-fluid theory provides only a first approximation, which is often satisfactory provided that the ratio of molecular-volume sizes does not exceed [about] 2. A much better approximation, however, has been given by Mansoori et al. [4].)

Alder's expression for Φ was extended to mixtures using an approximation similar to Equation 8.

To calculate vapor-liquid equilibria in a binary mixture, it is necessary to satisfy the relations

$$T^V = T^L \qquad (11)$$
$$P^V = P^L \qquad (12)$$
$$\mu_1^V = \mu_1^L \qquad (13)$$
$$\mu_2^V = \mu_2^L \qquad (14)$$

where superscripts V and L stand, respectively, for vapor phase and liquid phase. Pressure P is obtained from Equation 1 and chemical potential μ is obtained from Equation 2. Subscripts 1 and 2 stand, respectively, for methane and propane. To use Equation 2, we must remember, first, that mole fraction x_i is related to the total number of molecules N by

$$x_i = N_i/N \qquad (15)$$

and second, that the differentiation in Equation 2 is not performed at constant N but at constant N_j meaning that all N_j's (except N_i) are held constant. In that event, when N_i changes, N changes also.

Before Equations 11–14 can be used to obtain numerical results, it is necessary to obtain required molecular parameters from data reduction. For each pure fluid, we require two molecular parameters: a molecular size b and a molecular potential energy, usually designated by ϵ. These are obtained by fitting pure-component thermodynamic properties derived from the pure-component partition function, once for methane and once for propane. For vapor-liquid equilibrium calculations, it is usually best to obtain pure-component parameters by fitting experimental vapor pressures and experimental saturated liquid densities.

For the methane-propane mixture, only one adjustable binary parameter was used. This parameter is obtained upon fitting a few representative binary experimental data. For vapor-liquid equilibrium calculations, the most useful binary data are a few measurements of the compositions of coexisting (vapor and liquid) phases at a known temperature and pressure.

Simultaneous solution of Equations 11–14 is not simple, but computer programs are available for performing the trial-and-error calculations; a first estimate must be assumed. Equations 11–14 are then used to find a second estimate. Convergence is usually attained without difficulty, provided that the first estimate is reasonable.

Figure 3 shows calculated and experimental results for the methane-propane system. The results are conveniently expressed in terms of K factors defined by

$$K_i = y_i/x_i \tag{16}$$

where y is mole fraction in the vapor phase and x is mole fraction in the liquid phase.

Figure 3. Phase equilibria for methane-propane mixtures. Calculations based on a "modern" van der Waals equation of state.

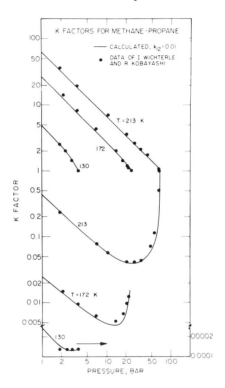

For this relatively simple binary mixture, good agreement with experiment was obtained using only two adjustable parameters for each pure fluid and one adjustable binary parameter, all independent of temperature, density, and composition. Similar results have been reported for many other mixtures of nonpolar nonelectrolytes with molecules that are not large (5). When molecules are large, it is nevertheless possible to use generalized van der Waals theory, but some phenomenological modifications are then required. One of these is described in the next section.

Generalized van der Waals theory for large molecules

For large molecules, and for fixed N, we cannot neglect the effect of density on rotational and vibrational degrees of freedom. To include contributions to the equation of state from rotational and vibrational degrees of freedom, it is convenient to factor these contributions into an internal part (a function only of temperature) and an external part, which depends primarily on density. For a pure fluid containing large molecules, the generalized van der Waals partition function is written

$$Q = \frac{1}{N!} \left[\frac{V_f}{\Lambda^3}\right]^N \left[\exp \frac{-\Phi}{2kT}\right]^N [q_{\text{ext } r,v}]^N [q_{\text{int } r,v}]^N \tag{17}$$

where $q_{\text{int } r,v}$ is independent of density and therefore does not contribute to the equation of state.

For the external part, we assume a simple relation (suggested by Donohue) which, in effect, is an interpolation between the known result at zero density and the high-density approximation of Prigogine, who assumed that external rotations and vibrations can be treated as equivalent translations (6).

Donohue suggests:

$$q_{\text{ext}\,r,v} = \left[\left(\frac{V_f}{V}\right)\left(\exp\frac{-\Phi}{2kT}\right)\right]^{c-1} \tag{18}$$

where $3c$ is the total number of external (density-dependent) degrees of freedom per molecule, i.e., all translations and all those molecular motions that are low-frequency, high-amplitude vibrations and rotations. For argon, $c = 1$, but for polyatomic fluids, $c > 1$. At liquid-like densities, where $V \approx$ constant, the partition function shown by Equations 17 and 18 is the same as that of Prigogine.

The domain of the generalized van der Waals partition function, here called perturbed-hard-chain (PHC) theory, is shown in Figure 4, where molecular complexity is plotted against density. PHC theory provides a smooth interpolation between well-understood boundary conditions: At low densities, we have theoretical results for simple (e.g., argon or methane) and complex (e.g., polyethylene) molecules.

Figure 4. Perturbed-hard-chain theory provides an equation of state for fluids containing simple or complex molecules, covering all fluid densities.

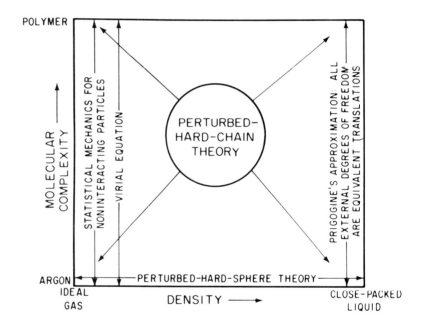

For simple molecules, we have theoretical results at all fluid densities (perturbed-hard-sphere theory); and at high densities, we have at least some theoretical knowledge if we adopt Prigogine's assumption that for liquids, all external degrees of freedom can be considered to be equivalent translational degrees of freedom. PHC theory is not rigorous, but it covers approximately a wide region of practical interest by meeting appropriate boundary conditions.

PHC theory can be reduced to practice by considering a large molecule to be the sum of molecular segments. Previously established expressions for V_f and for Φ for spheres are then used to give V_f and Φ for segments.

For example, suppose we use for V_f the simple expression originally given by van der Waals:

$$V_f = V - \frac{N}{N_A} b \tag{19}$$

For a system containing large molecules, each consisting of s segments, Equation 19 becomes

$$V_f = V - \frac{Nsb_s}{N_A} \tag{19a}$$

where b_s is the excluded volume per mole of segments. Equation 17 now becomes

$$Q = \frac{1}{N!} \frac{\Lambda^{-3N}}{V^{N(c-1)}} \left[\exp \frac{-\Phi}{2kT} \right]^{Nc} \left(V - \frac{Nsb_s}{N_A} \right)^{Ns(c/s)} (q_{\text{int } r,v})^N \tag{20}$$

where $3(c/s)$ is the number of external degrees of freedom per segment. Since a segment (by definition) is bonded to other segments, c/s is always less than unity when $s > 1$. Only when $s = 1$ is $c/s = 1$; in that event, every "segment" is not a segment but a free particle.

The exponent in Equation 20, therefore, reflects not only the number of segments Ns but also the relative freedom of these segments. In a sense, parameter c/s is a measure of overall segmental freedom normalized such that argon-like "segments" (where $c/s = 1$) have maximum freedom. A loose chain of segments (perhaps a rubber band) has a value of c/s larger than that for a stiff chain (perhaps a wooden rod) of the same length.

Extension to mixtures again follows from one-fluid theory. For each component, we require three adjustable molecular parameters: a segment-size parameter b_s; a segment-potential-energy parameter ϵ; and a segment-flexibility parameter c/s. From inspection of molecular structure we set an a priori value of s. For binary mixtures, we need one binary parameter, as suggested by Equation 9.

Figure 5. Henry's constants for volatile solutes in polyethylene.

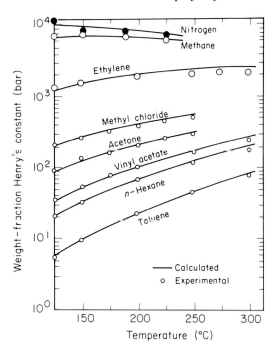

An application of PHC theory is shown in Figure 5, which correlates solubilities of volatile fluids in polyethylene over a broad range of temperature. The solubilities are expressed in terms of Henry's constant H, defined by

$$H = p/w \tag{21}$$

where p is the partial pressure of the solute and w is its weight fraction in polyethylene. The experimental solubility data were obtained using a gas-liquid chromatograph: The solid-support particles in the chromatographic column are coated with polyethylene. Helium flows slowly over the polymer-coated support. A small amount of solute is injected upstream of the column, along with a small amount of air (which is essentially insoluble in polyethylene). Downstream, the retention time of the solute is measured relative to that of air. This relative retention time is the key measurement for obtaining H (7).

Calculated results in Figure 5 are based on pure-component parameters obtained from volumetric data for polyethylene, and from volumetric and vapor-pressure data for the volatile solutes. Binary parameters were obtained from solubility data at one temperature.

Strongly nonideal liquid mixtures

Another application of molecular thermodynamics is directed at strongly nonideal liquid mixtures where orientational forces are appreciable—for example, mixtures of polar organic fluids, such as aldehydes, chlorinated hydrocarbons, ethers, nitriles, esters, and alcohols, including water. Because of strong preferential interactions arising from polarity and hydrogen bonding, it is not easily possible to describe such mixtures by a generalized van der Waals partition function coupled with the one-fluid theory of mixtures.

For liquid mixtures with strong orientational forces, we must take into account the tendency of molecules to segregate; that is, we must allow for the existence of local order where molecules do not mix at random, but instead show strong preferences in choosing their immediate neighbors. Although many outstanding scientists have tried to solve this difficult combinatorial problem, no satisfactory solution has been obtained for practical purposes. Therefore, it is necessary to construct an approximate model as suggested in Figure 6, which introduces the concept of local composition.

The essential idea here is that, because of local order, the composition in a very small region of the solution is not equal to the overall composition. To illustrate, Figure 6 shows 30 molecules, 15 shaded and 15 white; the overall mole fraction, therefore, is 1/2 for each component.

Figure 6. Local compositions and the concept of local mole fractions.

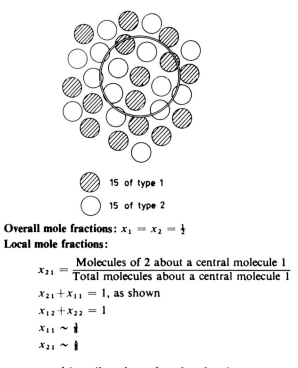

15 of type 1

15 of type 2

Overall mole fractions: $x_1 = x_2 = \frac{1}{2}$
Local mole fractions:

$$x_{21} = \frac{\text{Molecules of 2 about a central molecule 1}}{\text{Total molecules about a central molecule 1}}$$

$x_{21} + x_{11} = 1$, as shown

$x_{12} + x_{22} = 1$

$x_{11} \sim \frac{3}{8}$

$x_{21} \sim \frac{5}{8}$

We focus attention on one arbitrarily selected molecule of component 1 and describe the composition in the immediate vicinity of that molecule by two local mole fractions x_{11} and x_{21}; here x_{11} is the number of molecules of type 1 about the central molecule divided by the total number of molecules about that central molecule. An analogous definition holds for x_{21}, and there is the obvious conservation condition $x_{11} + x_{21} = 1$. As shown in Figure 6, there are eight molecules about the central molecule; five are white and three are shaded. Therefore, the local compositions are $x_{11} = 3/8$ and $x_{21} = 5/8$.

Although Figure 6 shows a molecule of type 1 at the center, a similar figure can be constructed with a molecule of type 2 at the center. For such a figure, we can define local mole fractions x_{22} and x_{12}.

The local-composition concept can be used to construct a simple expression for the nonideality of a liquid mixture, as shown in Figure 7. The important feature is the boxed equation, which expresses a two-fluid theory of binary mixtures: Molar internal energy U is given by the mole fraction average of $U^{(1)}$, the molar internal energy of hypothetical liquid 1, and $U^{(2)}$, the molar internal energy of hypothetical liquid 2. In hypothetical liquid 1, molecule 1 is at the center and the local mole fractions are x_{11} and x_{21}; similarly, in hypothetical liquid 2, molecule 2 is at the center and the local mole fractions are x_{22} and x_{12}. Parameter u_{ij} characterizes the potential energy of an ij pair; here $i = 1$ or 2 and $j = 1$ or 2.

Figure 7. Reduction to practice requires a model, classical thermodynamics, and a physical assumption suggested by statistical thermodynamics.

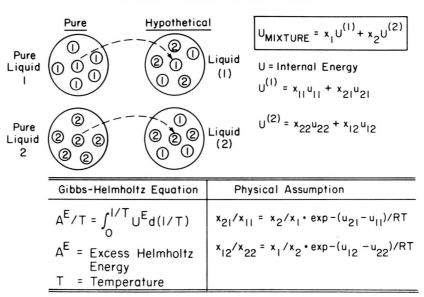

Given the relations shown in Figure 7, it is easily possible to derive an expression for ΔU, the energy change on mixing the real liquids ($\Delta U = U^E$), in terms of u_{ij} and the local mole fractions. This expression for ΔU is then substituted into the (classical thermodynamic) Gibbs-Helmholtz equation (shown at the lower left of Figure 7) to obtain an expression for the excess Helmholtz energy. From this we can obtain activity coefficients and thereby the deviations from ideality for the liquid mixture.

However, to carry out these calculations, it is necessary to compute the local mole fractions, which are conceptual quantities, not easily measured. To relate the conceptual local mole fractions to the measured overall mole fractions, we postulate a simple relation (shown at the lower right of Figure 7) suggested by statistical mechanics: For each hypothetical liquid, the ratio of local mole fractions is assumed to be equal to the ratio of overall mole fractions, multiplied by a Boltzmann factor. This relation is in no sense rigorous, but it provides a phenomenological approximation consistent with what statistical mechanics suggests about the local structure of an assembly of molecules that differ in intermolecular attractive forces.

Figure 7 illustrates some characteristic features of molecular thermodynamics for mixtures. At the lower left, we have classical thermodynamics; at the top, we have a simple physical model that attempts to reflect our limited knowledge of molecular behavior; and at the lower right, we have a relation suggested by statistical mechanics. Synthesis of these contributions to molecular thermodynamics yields an expression for calculating nonideality in a binary mixture in terms of two adjustable parameters $(u_{21} - u_{11})$ and $(u_{12} - u_{22})$.

For simplicity, Figures 6 and 7 represent only mixtures of molecules of equal sizes. However, the procedure outlined in Figures 6 and 7 can be generalized to molecules of different size by using local surface fractions instead of local mole fractions. Surface fractions are related to mole fractions using geometric properties of the molecules obtained from X-ray and similar data (8).

At low pressures, the molar excess Helmholtz energy A^E is essentially equal to the molar excess Gibbs energy G^E. Activity coefficient γ_i for component i is found by differentiation

$$\ln \gamma_i = \left(\frac{1}{RT}\right)\left(\frac{\partial nG^E}{\partial n_i}\right)_{T,P,n_j} \tag{22}$$

where n is the total number of moles and n_i is the number of moles of component i; R is the gas constant.

Phase equilibria are calculated using the equations of equilibrium. For liquid-liquid equilibria,

$$(\gamma_i x_i)' = (\gamma_i x_i)'' \qquad \text{for every } i \tag{23}$$

where ' and " designate the two liquid phases. For vapor-liquid equilibria,

$$\phi_i y_i P = \gamma_i x_i f_i^o \qquad \text{for every } i \tag{24}$$

where ϕ is the vapor-phase fugacity coefficient and f^o is the liquid-phase fugacity in the standard state, all at system temperature T and total pressure P. For typical liquid mixtures, f_i^o is the vapor pressure of pure liquid i.

For an ideal-gas mixture, $\phi_i = 1$. For real-gas mixtures, ϕ_i can be calculated from an equation of state.

Figure 8 shows a phase diagram for the system water-n-butanol at 1 atm. Calculations are based on Equations 23 and 24. For vapor-phase mixtures containing water and butanol at 1 atm, the ideal-gas assumption provides a satisfactory approximation.

Figure 8. Temperature-equilibrium phase composition diagram for butanol(1)-water(2) system. Calculations based on UNIQUAC equation with temperature-independent parameters.

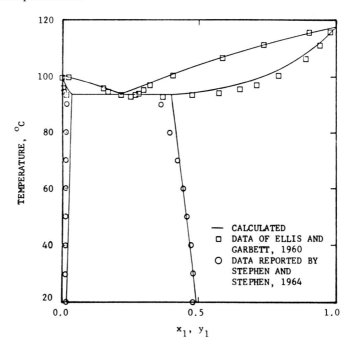

For the phase diagram shown in Figure 8, calculations for γ (Equation 22) use the UNIQUAC model, whose essential arguments are outlined in Figure 7, with one additional assumption concerning the high-temperature limit of the Gibbs-Helmholtz equation. The excess Helmholtz energy shown in Figure 7 is given by the indicated integral only if we assume that at very high temperature ($1/T \to 0$), the mixture is ideal (Raoult's law). UNIQUAC, however, makes a different assumption: It assumes that, as $1/T \to 0$, A^E is given by a modified Flory-Huggins expression for mixtures of (athermal) molecules that differ only in size. The modification of Flory-Huggins used here is that given by Staverman (9).

Figure 8 shows a temperature-composition diagram for a strongly nonideal mixture, based on the local-composition concept. By using only two adjustable parameters, it is possible to obtain an almost quantitative representation of the experimental data at constant pressure; at low temperatures, there are two liquid phases (partial miscibility), and at higher temperatures, the vapor-liquid equilibria exhibit an azeotrope.

Calculated results shown in Figure 8 are not in complete quantitative agreement with experiment; agreement could be much improved if adjustable parameters ($u_{21} - u_{11}$) and ($u_{12} - u_{22}$) were taken as linear functions of temperature. The remarkable feature of Figure 8 is that a simple theory, based on crude molecular approximations, can closely represent phase behavior of a highly complex liquid mixture using only two adjustable parameters.

Chemical effects on phase equilibria

The examples given above concern phase behavior in mixtures where molecules are chemically stable; i.e., there are no chemical reactions that accompany the mixing process. In many real mixtures, however, chemical equilibria have a strong influence on phase equilibria. To illustrate, consider the solubility of sulfur dioxide in water at low pressure. Normally, the low-pressure solubility of a gas in a liquid is given by Henry's law, where the partial pressure of the solute is proportional to its concentration in the liquid phase. However, when sulfur dioxide dissolves in water, a chemical reaction occurs: Aqueous sulfur dioxide dissociates into bisulfite ion and hydrogen ion. Thus chemical equilibrium (dissociation) must be superimposed on phase equilibrium (Henry's law).

The first column in Table 2 gives the partial pressure of sulfur dioxide. The second column gives the molality of sulfur dioxide in aqueous solution. When these data are plotted, they do not yield a straight line; despite the very low concentrations of sulfur dioxide, Henry's law is not valid.

Table II. Solubility of sulfur dioxide in water at 25 °C

Partial pressure, p_{SO_2} (bar)	Molality, m (mol SO_2/1000 g H_2O)	Fraction ionized, α	Molality of molecular SO_2, m_M (mol SO_2/1000 g H_2O)
0.0105	0.0271	0.524	0.0129
0.0456	0.0854	0.363	0.0544
0.0984	0.1663	0.285	0.1189
0.1811	0.2873	0.230	0.2212
0.3374	0.5014	0.184	0.4092
0.5330	0.7643	0.154	0.6470
0.7326	1.0273	0.134	0.8897
0.9312	1.290	0.120	1.134
1.0822	1.496	0.116	1.329

Source: Reference 10.

The failure of Henry's law becomes clear when we consider the consequences of ionization. When sulfur dioxide dissolves in liquid water, we must take into account two equilibria, as shown by the vertical and horizontal arrows:

Gas phase	SO_2
Liquid phase	SO_2 (aqueous) $\rightleftharpoons H^+ + HSO_3^-$

Henry's law governs only the (vertical) equilibrium between the two phases; in this case Henry's law must be written

$$p = Hm_M \tag{25}$$

where p is the partial pressure of sulfur dioxide, H is a "true" Henry's constant, and m_M is the molality of molecular (non-ionized) sulfur dioxide in aqueous solution.

Johnstone and Leppla (10) have calculated α, the fractional ionization of sulfur dioxide in water, from electrical conductivity measurements; these are given in column 3 of Table 2. Column 4 gives m_M, which is the product of m, the total molality, and $(1 - \alpha)$. When the partial pressure of sulfur dioxide is plotted against m_M (rather than m), a straight line is obtained.

This case is a particularly fortunate one because independent conductivity measurements are available; thus it is possible quantitatively to reconcile the solubility data with Henry's law. In a more typical case, independent data on the liquid solution (other than the solubility data themselves) would not be available; however, even then it is possible to linearize the solubility data by constructing a simple but reasonable model.

For equilibrium between sulfur dioxide in the gas phase and molecular sulfur dioxide in the liquid phase, we write

$$p = Hm_M = Hm(1 - \alpha). \tag{26}$$

For the ionization equilibrium in the liquid phase, we write

$$K = \frac{m_{H^+} m_{HSO_3^-}}{m_M} = \frac{\alpha^2 m^2}{m_M} \tag{27}$$

where K is the ionization equilibrium constant. Substituting Equation 26, we have

$$K = \frac{\alpha^2 m^2}{p/H} \tag{28}$$

from which we obtain α:

$$\alpha = \frac{\sqrt{p}}{m} \left(\frac{K}{H}\right)^{1/2} \tag{29}$$

Further substitution and rearrangement finally give

$$\frac{m}{\sqrt{p}} = \frac{\sqrt{p}}{H} + \left(\frac{K}{H}\right)^{1/2} \tag{30}$$

Equation 30 shows the effect of ionization on Henry's law; if there is no ionization, $K = 0$ and Henry's law is recovered. The tendency of a solute to ionize in solution increases its solubility; however, as the concentration of solute in the solvent rises, the fraction ionized falls. Therefore, the "effective" Henry's constant p/m rises with increasing partial pressure; a plot of p versus m is not linear but convex toward the horizontal axis.

Figure 9 presents solubility data for the sulfur dioxide/water system plotted according to Equation 30. A straight line is obtained. A plot like this is useful for

Figure 9. Solubility of sulfur dioxide in water at 25 °C. Linear plot follows from taking ionization into account.

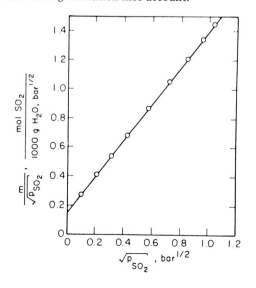

smoothing, interpolating, and cautiously extrapolating limited gas solubility data for any system where the gaseous solute tends to ionize (or dissociate) in the solvent.

Conclusion

The few examples cited here illustrate how typical phase equilibria can be quantitatively described using the principles of molecular thermodynamics. When well-known concepts from statistical thermodynamics and chemical physics are joined with classical thermodynamics, it is often possible to establish an approximate model that captures the essential phenomena with sufficient accuracy to provide a quantitative (or nearly quantitative) description. Such models inevitably contain one or two empirical parameters that must be obtained from a few representative experimental measurements.

Molecular thermodynamics cannot eliminate the need for experimental data to achieve a description of phase equilibria in real systems. However, molecular thermodynamics very much reduces that need. Molecular thermodynamics is efficient in the sense that, when physical chemistry is used with imagination, it can often describe a variety of complex equilibrium phenomena with little experimental effort.

The author is grateful to the National Science Foundation for financial support and to Edmundo G. de Azevedo for his review of the manuscript.

References

(1) Hill, T. L. *An Introduction to Statistical Mechanics;* Addison-Wesley: New York, 1960.
(2) McQuarrie, D. A. *Statistical Mechanics;* Harper & Row: New York, 1976.
(3) Alder, B. J.; Young, D. A.; Mark, M. A. *J. Chem. Phys.* **1972**, *56*, 3013.
(4) Mansoori, G. A.; Carnahan, N. F.; Starling, K. E.; Leland, T. W. *J. Chem. Phys.* **1971**, *54*, 1523.
(5) Knapp, H.; Döring, R.; Oellrich, L.; Plöcker, U.; Prausnitz, J. M. *Vapor-Liquid Equilibria for Mixtures of Low Boiling Substances;* Dechema Chemistry Data Series: Frankfurt, 1982.
(6) Prigogine, I. *The Molecular Theory of Solutions;* North-Holland Publishing: Amsterdam, 1957.
(7) Smidsrod, O.; Guillet, J. E. *Macromolecules* **1969**, *2*, 272; Newman, R. D.; Prausnitz, J. M. *J. Phys. Chem.* **1972**, *76*, 1492.
(8) Bondi, A. *Physical Properties of Molecular Crystals, Liquids and Glasses;* John Wiley & Sons: New York, 1968.
(9) Staverman, A. J. *Recl. Trav. Chim. Pays-Bas* **1950**, *69*, 163.
(10) Johnstone, H. F.; Leppla, P. W. *J. Am. Chem. Soc.* **1934**, *56*, 2233; Campbell, W. B.; Maass, O. *Can. J. Res.* **1930**, *2*, 42; Morgan, O. M.; Maass, O. *Can. J. Res.* **1931**, *5*, 162.

Chapter 7
POLYMER EXAMPLES OF THERMODYNAMICS, STATISTICAL MECHANICS, AND CHEMICAL KINETICS

LEO MANDELKERN
Department of Chemistry and Institute of Molecular Biophysics
Florida State University
Tallahassee, Fla. 32306

Natural and synthetic polymers are widely observed in our universe and are major constituents of our everyday life. Yet in most situations, students are rarely offered the opportunity to learn about this class of molecules. There are several reasons for this distressing state of affairs. One reason is clearly historical (1). The basic underlying pattern of chemical education, whose essential form has remained unaltered for many decades, was established prior to the recognition of a rigorous, as well as an extremely vigorous, polymer or macromolecular science. This unfortunate accident of history has not as yet been rectified. Another reason for the lack of opportunity to study polymers is the widespread concept that polymers are very different, are very complex, and require their own unique and peculiar laws of nature. The late Professor Paul J. Flory has pointed out quite clearly that "the frequent assertion that macromolecules are intrinsically too complicated, and therefore too difficult, to be comprehended by undergraduates can be controverted..." (1).

I hope that the examples that will be presented in subsequent sections of this report will make this point quite clear and will help set aside the dogma that has unfortunately pervaded chemical education for much too long a time. We shall in fact find that the study of small and large molecules is a continuous process. There is no defined boundary separating one class from the other. When examined in retrospect, it is apparent that the study of polymers has benefited from the knowledge of small molecules. On the other hand, investigations of polymer systems have enriched and generalized our knowledge of the chemistry of small molecules. In the development of polymer science, some very important scientific principles have been established that are applicable to all areas of chemistry. They have led to a keener insight and a more general understanding of molecules of all sizes.

The extreme technological importance of polymers in commerce and in biology notwithstanding—and there is no question that they are very important—macromolecules are very interesting molecular species in their own right. The supposed complexities are in fact nonexistent. As we shall see, no new laws of nature have to be formulated to explain their properties and behavior. No new principles need to be involved. The introduction of polymer concepts enhances the undergraduate curriculum and arouses and holds students' interest. Indeed, it can be asserted that as much pedagogical information can be transmitted by using macromolecular systems as examples as by using gases and simple liquids. The major purpose of this report is to demonstrate, by means of definitive examples, how polymer principles can be introduced in a very natural way into the first undergraduate physical chemistry course without sacrificing any of the rigor and mental discipline that have traditionally, and very properly, been expected.

Professor John D. Ferry (2) pointed out some 25 years ago that the then new material of polymer chemistry could be introduced to illustrate basic concepts without departing from the customary development of the subject. The study of polymers should and can be integrated into the logical exposition of the various subdisciplines of physical chemistry. Polymers should not be set aside as a separate, isolated entity since they can be woven into the fabric of physical chemistry. None of the traditional subjects of physical chemistry need to be neglected in this undertaking. One does not have to be a practitioner of polymer science to be able to introduce the concepts into the undergraduate curriculum. When properly chosen, the examples will demonstrate that polymers do not need special study, nor do they require their own laws. In fact, very important generalizations result that have enhanced our understanding of the physical chemistry of all types of molecular systems.

The influence of polymers is felt in all facets of physical chemistry. The most striking examples are found in many aspects of thermodynamics, in statistical mechanics, in conformational analysis, and in chemical kinetics and catalysis. As polymer structures are being investigated and probed in more detail by nuclear magnetic resonance and other spectroscopic techniques, influences in the teaching of quantum mechanics are beginning to be felt.

Major efforts have already been made to set forth specific examples that could be introduced into the first physical chemistry course (2–8). I hope to enhance and expand upon these concepts in the sections that follow by detailing examples in some of these major areas. It should be recognized that this chapter leans very heavily on the report of the Education Committee of the Polymer Division of the American Chemical Society, which has been published in the *Journal of Chemical Education* (8).

Thermodynamics

Polymers can be very useful tools in expanding the vistas and use of thermodynamics. The concepts are more viable and palatable as students become involved in a class of substances to which they can relate very comfortably. The question is simply whether it is more interesting for a student to become involved with the properties of a rubber band or with an ideal gas. The thermodynamic laws are obviously the same for both.

Several distinct features of long-chain molecules manifest themselves in the thermodynamic laws. One of these results from the fact that in dealing with polymers one immediately and intuitively needs to consider a complete set of intensive-extensive variables, rather than just the P-V pair, as is usually presented. This approach quickly leads to a generalization of many principles that become obvious in retrospect. Another major set of applications has its roots in the unique configurational versatility of disordered polymer chains. This property manifests itself in very striking ways in solution thermodynamics, in problems ranging from ideal solutions to liquid-liquid phase separation. In the next sections are described some examples that can be integrated in a straightforward manner into the first physical chemistry course.

Thermochemistry

Despite the fact that the correct conditions are invariably stated, the idea that $\Delta H = q_p$ (the enthalpy change is equal to the heat absorbed at constant pressure) is the one that students usually retain. This situation arises because the examples to which they are exposed obey this relation. This concept is of course limited to processes that only involve P-V as the work variables. However, many reactions, particularly some very important ones of biological and biochemical interest, take place on membranes and fibers. In these cases the enthalpy change does not equal q_p. Put another way, for these types of reactions the enthalpy tables do not directly yield the heat change. This point is not always recognized. Thus by considering reactions on fibers or membranes, the thermochemistry laws can be presented in a very general manner.

Carnot cycle

The Carnot cycle is a very useful concept and pedagogical tool that provides a definitive example of the limitation of the First Law and serves as a good introduction to the Second Law. Almost universally, in textbooks and classroom examples, an ideal gas is used as a working substance because its equation of state allows a simple analysis to be made. Then, either it is asserted or the theorem proved that the working substance need not be an ideal gas. This generalization can be made quite vivid by using a rubber band as the working substance in force-length space. A closed loop is obtained because the adiabatics are steeper than the isothermals (9, p. 209 ff.; 10). Although this system is somewhat more difficult to analyze quantitatively than an ideal gas, a diagrammatic representation of a cycle, or engine, can be constructed with a rubber band. The subsequent discussion removes the concept from the abstract and makes it more interesting. In fact, a simple engine working on this principle, the *Wiegand pendulum*, can be constructed from a rubber band (11).

Second Law

The Second Law of Thermodynamics and its consequences are usually quite difficult for the beginning student to grasp. The consequences of the Second Law can be dramatically illustrated by the properties of rubber-like substances using nothing more complex than a rubber band. In 1806, well before the enunciation of the Second Law, Gough reported (12) that caoutchouc (natural rubber) possessed thermal and elastic properties that were dramatically different from those of all other substances. He found that when a rubber band is stretched rapidly, i.e., adiabatically, its temperature rises. Sophisticated measurements have since shown that there is about a 10 °C temperature rise for a four- to five-fold extension. The experiment is easily demonstrated, and it can be performed by students, by rapidly expanding an ordinary rubber band and immediately placing it to their lips. Shortly after the enunciation of the Second Law it was deduced that for this substance, unique among all those studied, the entropy must decrease with extension.

This observation, coupled with the Second Law, predicts a unique thermal expansion coefficient for a rubber-like substance under stress. This expectation is indeed fulfilled and is easily demonstrated for the class. As is schematically illustrated in Figure 1, a positive thermal expansion coefficient is displayed by virtually all low molecular weight systems as well as by undeformed polymers. However, stretched rubber possesses a negative thermal expansion coefficient in the direction of deformation; i.e., it contracts on heating and expands on cooling. The two observations described are consistent with one another, as can be shown by the analysis of the Gibbs Free Energy (see below). These examples also illustrate the relationship between thermodynamic formalism and macroscopic properties well beyond the concept of the expansion or compression of an ideal gas.

Figure 1. Thermal expansion of a rubber-like substance (a) unstressed, (b) stressed. *Source*: Reference 13.

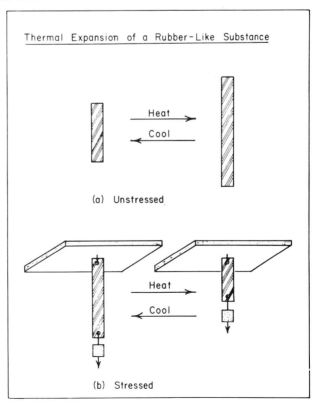

Gibbs Free Energy

It is instructive to consider the change in the Gibbs Free Energy, G, during the deformation of a rubber-like substance. The intensive-extensive variables force (f) and length (L) need to be introduced into the problem. It then follows that

$$dG = Vdp - SdT + fdL \qquad (1)$$

Upon applying Maxwell's relation to f and S one obtains

$$f = \left(\frac{\partial H}{\partial L}\right)_{T,p} + T\left(\frac{\partial f}{\partial T}\right)_{p,L} \qquad (2)$$

Equation 2 is the thermodynamic equation of state for elasticity and is completely analogous to the equation of state of a gas. The comparison can be carried further for, in analogy to an ideal gas, an ideal rubber could be defined as one where

$$\left(\frac{\partial H}{\partial L}\right)_{T,p} = 0$$

The unique thermal properties cited above follow from the general properties of partial derivatives and Maxwell's relation. Thus

$$\left(\frac{\partial f}{\partial T}\right)_{p,L} = -\left(\frac{\partial S}{\partial L}\right)_{T,p} = -\left(\frac{\partial f}{\partial L}\right)_{T,p}\left(\frac{\partial L}{\partial T}\right)_{f,p} \qquad (3)$$

Since the modulus $(\partial f/\partial L)_{T,p}$ of an elastic body is positive by definition, a negative thermal expansion coefficient must be observed if the entropy decreases with an increase in length.

Besides the formalisms that are involved in the above analysis, an insight into the molecular origin of molecular properties can be developed. The observation of a unique thermal expansion coefficient and the required decrease in entropy with an increase in length lead to an inquiry as to the kind of structures that may be involved. Since this behavior is unique to polymeric systems, the general configurational features of long-chain molecules in disordered conformations can be examined (even in a very qualitative way) prior to a more detailed statistical mechanical analysis.

Phase Equilibria

Polymers can be used in many interesting examples involving problems in phase equilibria. We can start with the phase rule, which can be expressed in its most general form as

$$F = C - P + I \qquad (4)$$

Here C and P have their usual meanings and I is the total number of intensive variables that describe the system. In the usual presentation, since only T and P are considered to be the intensive variables, $I = 2$. The modification of Equation 4, with $I = 2$, is so ingrained that the impression has been created that it represents the phase rule. The general form of Equation 4 is very useful in analyzing problems concerned with polymeric fibers and membranes, particularly if phase transitions are involved (*14*).

The discussion of solid-liquid phase equilibria can be illustrated and generalized by examples taken from the polymer literature. Many such examples demonstrate that for isotropic systems, the conventional Clapeyron equation is obeyed with the usual P-V variables (*15, 16*). However, a more interesting situation arises when a polymeric system is subject to a uniform tensile force at constant pressure. When equilibrium is maintained between two phases of a pure species, such as the crystalline phase (fiber) and liquid phase (rubber), it immediately follows that (*9,* p. 170 ff.; *17, 18*)

$$\left(\frac{\partial f}{\partial T}\right)_{p,\text{eq}} = -\frac{\Delta S}{\Delta L} \qquad (5)$$

Here S and L represent the changes in the entropy and length that take place upon the transformation of the entire fiber, from one phase to the other, at constant T and P. Since

$$\Delta S = \frac{\Delta H - f\Delta L}{T} \qquad (6)$$

(note the additional term in the numerator), it follows that

$$\left[\frac{\partial (f/T)}{\partial (1/T)}\right]_{p,\mathrm{eq}} = \frac{\Delta H}{\Delta L} \tag{7}$$

The analogy of Equation 7 to the conventional Clapeyron equation is clear when it is recognized that there is a correspondence of $-f$ with P, and V with L. Coexistence curves between the liquid and crystalline phases can be determined experimentally. These are analogous to those obtained for vapor-liquid equilibria. As an example, we give in Figure 2 a plot of the tension-length relation for fibrous natural rubber,

Figure 2. Composite plot of tension-length relation at various temperatures for fibrous natural rubber. *Source*: Reference 19.

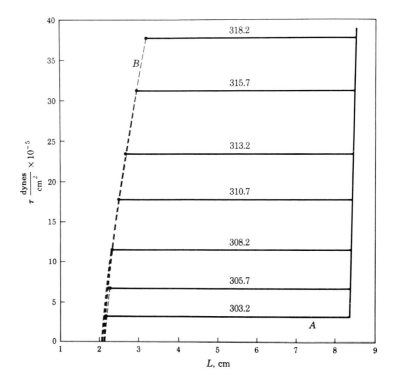

determined at various temperatures through the two-phase region (*19*). The horizontal solid lines represent the stresses necessary to maintain the two phases in equilibrium. As required by the phase rule, and in analogy to the coexistence curves representing vapor-liquid equilibria, the force is found to be independent of length in the two-phase region. The solid circles represent the length of the sample in the liquid state as two-phase equilibrium is developed, i.e., at the melting temperature T_m of the sample. The dashed lines represent the force-length relations at each temperature in the liquid state as calculated from rubber elasticity theory (see below). The force-length relation in the crystalline state at 303.2 K is represented by the vertically rising straight lines. The equation of state at other temperatures in the crystalline state has a similar representation. The correspondence of the coexistence curves in Figure 2 to those describing vapor-liquid equilibria of nonpolymeric substances is readily apparent when the relationships between the respective intensive-extensive variables are recognized.

As a consequence of the transition that is involved in the systems described above, large tension can be developed at constant length. Conversely, anisotropic dimensional changes can take place at constant force. Therefore, very simply constituted fibrous systems can serve as, and demonstrate, reversible contractile systems. A more detailed examination of Figure 2 illustrates the relations that exist between the force, length, and temperature through the two-phase region. For example, consider the system of Figure 2 at point A. If the temperature is increased while the length remains constant, a path described by a vertical line upward will be followed. For the length to remain constant under these conditions, an external force must be applied. This additional stress is needed to prevent melting as the temperature is increased. Otherwise, the original length could not be preserved. If the temperature is raised to 318.2 K (the highest temperature studied in these particular experiments), a tensile force of 4 kg/cm^2 will be developed. This tension is the same order of magnitude as is developed by the muscle fiber system when in tetanic contraction. The stress will continuously increase with temperature, as long as the two-phase region is maintained, until the critical temperature is reached.

One can also consider a process where the stress rather than the length is held constant as the temperature is changed. If we start again at point A of Figure 2, as the temperature is raised, a horizontal path will be followed that terminates at the appropriate dashed curve representing the liquid or rubber-like state. Accompanying this transformation at constant stress will be a fourfold decrease in the length of the sample.

It can thus be demonstrated that the melting-crystallization process of polymeric systems under equilibrium conditions can lead to large reversible dimensional changes when carried out at fixed force, or, conversely, very large tensions can be developed when carried out at constant length. The analysis for the one-component system that has been described can be generalized to multicomponent systems. Situations where direct chemical interaction occurs between the polymer and low molecular weight species can also be accounted for (*9*, p. 182 ff.; *20, 21*). With this generalization, coexistence curves in the *f-L* plane can be constructed at constant temperature with variation taking place in the chemical potentials of any of the species in any of the states as a consequence of concentration changes or specific interactions. Therefore, contractile-tension developing systems can also be developed isothermally.

The tension and dimensional changes that have been described will occur very sharply, with only small changes in temperature or chemical potential, because the process has all the characteristics associated with classical first-order phase transitions (see below). This mechanism for tension development and shortening can serve as a basis for understanding naturally occurring contractile processes (*18, 19, 22*) as well as for the development of synthetic mechanochemical devices (*23, 24*). These are very important principles that are applicable to many biochemical phenomena and to the important societal problems involving the conversion of free energy sources into useful work. These matters are rarely, if ever, examined in an undergraduate physical chemistry course. The use of the polymeric examples cited above allows these problems to be addressed, and brought into the classroom, in a very straightforward manner.

The discussion of the Clapeyron equation and its ramifications implied that the melting-crystallization process involving polymers can be treated in a classical manner. Despite the structural and morphological complexities characteristic of the crystalline state, it has been shown theoretically and experimentally that the melting-crystallization of polymers can be treated as a first-order phase transition, although the fusion process is somewhat diffuse (*9*, p. 20 ff.; *25*). This conclusion is based on theoretical analysis supported by experimental studies of the influence of chain length, pressure, and diluent on the melting temperature (*9*, p. 20 ff.). The influences of these factors are predictable by the usual thermodynamic relations appropriate to a first-order phase transition.

The fusion of isotropic polymer systems (i.e., in the absence of any applied external force) provides many examples of structural influences on the fusion process and leads to a more general understanding of the process. As one example, the chemical purity of a one-component system can be judged by the sharpness of melting and the depression of the melting temperature. However, "morphological impurities" such as surface effects, finite crystallite sizes, and imperfection within the crystalline state also cause melting-point depressions and a broadening of the fusion range. These effects can be observed in all types of crystalline substances but are particularly manifest in polymeric systems. Here relatively small crystallite sizes and high interfacial energies are quite common. However, it is also possible under careful crystallization conditions to develop a relatively sharp melting system. This point is illustrated in Figure 3, where the fusion of two different, but chemically identical, linear polyethylene samples is illustrated. In both of these examples the melting temperatures are well defined. For the molecular weight fraction, the fusion process is relatively sharp. In contrast, for the same chemically pure but polydisperse polymer there is a substantial broadening of the fusion range caused by the kinds of morphological impurities cited above.

Figure 3. Specific volume-temperature relations for linear polyethylene samples. Solid circles, unfractionated polymer; open circles, fraction. $M = 32{,}000$. *Source*: Reference 26.

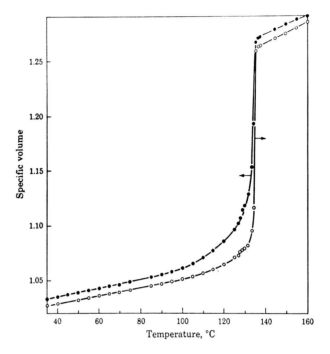

In the undergraduate laboratory, fluidity observations, usually made in a capillary tube, are taken as a measure of melting. Melting is, of course, a thermodynamic phenomenon. However, as long as the liquid state has a very low viscosity there is no problem in determining melting points in this manner. For high-viscosity systems, exemplified by polymers as well as complex low molecular weight substances, the capillary method fails and direct thermodynamic methods must be used. Very serious error can be made by not using direct thermodynamic methods. In fact, in some instances involving very high molecular weight polymers, melting is not recognized by the capillary method.

Distinct chemical impurities influence the fusion of polymers, as they do all substances. They do so in some very interesting and unique ways. In this discussion, we must distinguish between an added second component, or diluent, and structural irregularities directly incorporated into the chain, which can also act as impurities. Analysis of the former case proceeds in standard fashion. The distribution of the component between the phases must be stated a priori. The chemical potentials of the species in each phase are then equated, with the melting point-composition relation resulting. In the most common case, where the crystalline phase remains pure, using the Flory-Huggins equation (*27; 28; 29*, Ch. XII) for the free energy of mixing in the liquid state, the melting temperature relation that results can be expressed as (*25; 29*, Ch. XIII)

$$\frac{1}{T_m} - \frac{1}{T_m^\circ} = \left(\frac{R}{\Delta H_u}\right)\left(\frac{V_u}{V_1}\right)[(1 - v_2) - \chi_1(1 - v_2)^2] \qquad (8)$$

Here T_m° is the melting temperature of the pure polymer, and T_m is the melting temperature of the polymer-diluent mixture whose composition is specified by the volume fraction v_2 of polymer. ΔH_u represents the heat of fusion per polymer repeating unit, χ_1 the polymer-diluent interaction parameter (*30*); V_u is the molar volume of the repeating unit and V_1 that of the diluent. Equation 8 is a generalized form of the traditional freezing-point depression equation, applied in this case to long-chain molecules. The colligative nature of the expression is apparent, and it has been quantitatively substantiated by a large amount of experimental data (9, p. 20 ff.).

Unique to polymers, however, is the situation where the impurities are actually built into the chain. These include chain ends (which do not usually enter the crystal lattice), chemically different units, and more subtle effects such as geometric and stereoisomers, branch points, and head-to-head or tail-to-tail structures. Quantitative analysis has been made for most of these possibilities. For example, for a homopolymer having a most probable molecular weight distribution (25),

$$\frac{1}{T_m} - \frac{1}{T_m^\circ} = \frac{R}{\Delta H_u} \frac{2}{\bar{x}_n} \tag{9}$$

Here T_m° is the melting temperature of the infinite molecular weight chain, T_m the melting temperature of the system, and $\bar{x}_n$ the number average degree of polymerization. Equation 9 represents a very general effect of noncrystallizing species, as there are two ends per molecule whose average length is given by $\bar{x}_n$. The colligative nature of the melting-point depression is also illustrated by Equation 9 since as $\bar{x}_n$ increases, the melting-point depression becomes smaller.

The other types of impurities that are an inherent part of the chain can be grouped together in the generic category of copolymers. These represent an interesting case, which though unique to chain molecules has important and very broad general implications. If we consider a chain composed of A and B units (which are either chemically or structurally different) and only allow the A units to crystallize, then it has been shown by a straightforward analysis that (25, 30)

$$\frac{1}{T_m} - \frac{1}{T_m^\circ} = \frac{-R}{\Delta H_u} \ln p \tag{10}$$

Here T_m is the melting temperature of the copolymer and T_m° is the melting temperature of the corresponding homopolymer. The quantity p is defined as the sequence propagation probability. This quantity represents the probability that an A unit selected at random in a chain is succeeded by another A unit. This result is obviously a unique property of chain molecules. It predicts that the melting temperature, and thus related properties, will depend on how the sequences are arranged, rather than directly on composition. In addition, the specific chemical or structural nature of the co-unit is not important by itself—admittedly a rather unusual result. For example, for a random copolymer, $p = X_A$, the mole fraction of A units; for an ordered or block copolymer p will approach unity; for an alternating copolymer p will be less than X_A. The striking prediction that widely varying melting temperatures can be achieved with copolymers of the same composition has received widespread experimental verification (9, p. 82 ff.)

A striking, but typical, set of results, which is well suited for classroom presentation, is given in Figure 4 for a series of copolyesters based on ethylene terephthalate (31). (The homopolymer, polyethylene terephthalate, can be recognized by its common name Dacron or Mylar.) Here, the different chemical co-units are arranged either randomly or in blocks with the ethylene terephthalate. The random copolymers, independent of the nature of the co-unit, show a severe melting-point depression that can be as much as 200 °C. On the other hand, the melting temperatures of the ordered copolymer fulfill the theoretical prediction in that they are essentially invariant with composition. Many practical uses of such results take advantage of the fact that composition can be maintained while crystallinity and related properties can be carefully controlled.

An analysis of the data that have been obtained in studying the melting of polymers offers the opportunity to focus on the relationship between the entropy of fusion, molecular structure, and the role of the entropy of fusion in determining the location of the melting temperature. It is found that the entropy of fusion correlates quite well with the conformational properties of the disordered chain (9, p. 113 ff.; 32). With almost no exception, high-melting polymers possess low entropies of fusion, while the low-melting polymers have very high entropies of fusion.

Solution Properties

The dilute solution behavior of both flexible and rigid long-chain molecules offers many different and interesting examples that are suitable for classroom discussion. It is convenient to start with the concept of an ideal solution, which is classically defined as

$$\Delta H_M = 0$$

$$\Delta V_M = 0$$

$$\Delta S_M = -R \sum_i n_i \ln x_i \qquad (11)$$

Here n_i and x_i are the number of moles and mole fraction of species i, whereas ΔH_M, ΔV_M, and ΔS_M are the enthalpy, volume, and entropy of mixing, respectively. These conditions immediately lead to Raoult's law, the usual way of expressing ideality. The conditions that ΔH_M and $\Delta V_M = 0$ imply no net interaction between the species, an obvious requirement for ideality. However, a very simple statistical derivation, using lattice methods, shows that the expression for ΔS_M requires the species to be of equal size (29, Ch. XII). Therefore, when expressing ideality by means of Raoult's law, there are two restraints; one involves interactions, the other size. However, when the ideal entropy of mixing of flexible long-chain molecules is calculated by similar methods, as in the theories of Huggins (27) and Flory (28; 29, Ch. XII), it is found that

$$\Delta S_M = -R \sum_i n_i \ln v_i \qquad (12)$$

Here v_i is the volume fraction (as contrasted with the mole fraction of Equation 11). Equation 12 reduces to Equation 11 when the species in the mixture are of equal size. Thus the conventional criteria are regenerated when the size requirement is fulfilled. However, Equation 12 can serve as a much more general expression for ideality, as the interaction factors are isolated from the size requirements. Equation 12 has been a very good base for understanding polymer solutions. It allows for a rational comparison to be made of the properties of real solutions in the situation where the

Figure 4. Melting temperature-composition relations for copolymers of ethylene terephthalate. Ordered copolymers with (1) ethylene succinate; (2) ethylene adipate; (3) diethylene adipate; (4) ethylene azelate; (5) ethylene sebacate; (6) ethylene phthalate; and (7) ethylene isophthalate. Random copolymers with (2) ethylene adipate and (5) ethylene sebacate. *Source*: Reference 31.

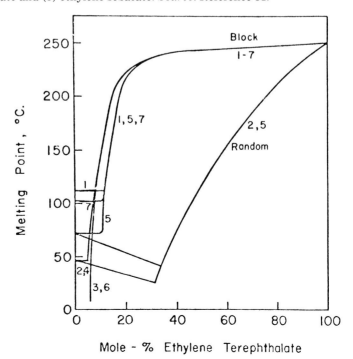

classical Raoult's law fails (29, Ch. XII). The analysis of the thermodynamics of polymer solutions has improved our understanding of the basis of Raoult's law and its implications. A very general formulation of ideality has evolved, which encompasses species of all sizes and which is very potent in analyzing solution behavior.

In dilute solution it is convenient to express the solvent activity, for any type solute, in virial form. In the case of the osmotic pressure, for example, this formalism is analogous to expressing PV of a real gas in powers of $1/V$. Following this procedure, the osmotic pressure, π, can be expressed as

$$\frac{\pi}{c} = RT(A_1 + A_2 c + \ldots) \tag{13}$$

Here c is the solute concentration and the A's are the virial coefficients. A_1, the first virial coefficient, is equal to $1/M$, the reciprocal of the molecular weight. The second virial coefficient is expressed as

$$A_2 = \frac{Nu}{2M^2} \tag{14}$$

Here u is the intermolecular excluded volume and N is Avogadro's number. The value of u has been calculated for different shaped macromolecules in different thermodynamic environments (29, Ch. XII). When $u = 0$, the second virial coefficient vanishes, the solution becomes ideal, and Equation 13 becomes the van't Hoff expression for the osmotic pressure.

For many solutions of flexible chain molecules a temperature $T = \theta$ can be found at which A_2 vanishes and the solution behaves ideally (29, Ch. XII). This condition corresponds to the Boyle point of a real gas, where, for a very specific T and P, the ideal equation of state is obeyed. An example of this behavior for polymer solutions is illustrated in Figure 5, where π/c is plotted against c for four polyisobutylene fractions in benzene at 20 °C, 30 °C, and 40 °C (33). It is clear in these plots that the

Figure 5. Plot of π/c against concentration c of molecular weight fractions of polyisobutylene at indicated temperatures. LB3 $M_N = 1.01 \times 10^5$; LA1 $M_N = 1.91 \times 10^5$; LI-345-5 $M_N = 2.06 \times 10^5$; PAB1F $M_N = 7.1 \times 10^5$. Source: Reference 33.

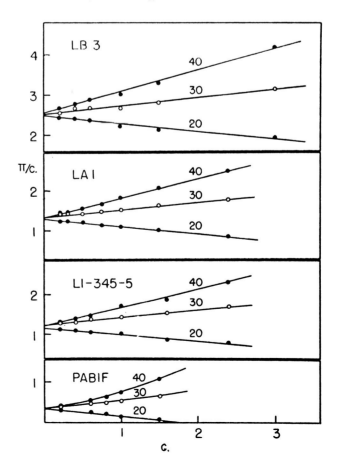

initial slope, which is directly related to A_2, changes sign as the temperature is lowered. Therefore, at some temperature between 20 °C and 30 °C, A_2, as well as the higher virial coefficients, will vanish and the solution will become ideal. Thus, these kinds of polymer solutions demonstrate ideality quite vividly over any extended temperature range. However, it should be noted that at elevated temperatures, where the solution is no longer ideal, $\pi/c - c$ plots deviate from linearity at very low concentrations. Therefore, as a very general proposition, the point can be made that one must be wary both of linear extrapolations and of the unsupported statement that one is prone to make, that a solution is sufficiently dilute so that Raoult's law is obeyed.

Colligative Properties

Measurements on polymer solutions can dramatically illustrate the concept of colligative properties and very forcefully point out the limitations that exist in molecular weight determination. Since these types of measurements count the number of molecules, the boiling-point elevation, freezing-point depression, and osmotic pressure will decrease as the molecular weight increases at a fixed solute concentration. For even modest molecular weights, $M \simeq 10^4$, changes in melting and boiling points are only of the order of 10^{-3} °C. Therefore, changes in these quantities will become progressively more difficult to detect and will require extraordinary experimental techniques. On the other hand, although the osmotic pressure is also inversely proportional to the molecular weight, the sensitivity of the method is such that very high molecular weights can be determined. For example, the highest molecular weight illustrated in Figure 5 is 7×10^5. With proper care, even higher molecular weights can be determined. At the other extreme, solutes as small as sucrose can also be measured.

Liquid-Liquid Phase Separation

Under appropriate conditions, polymer solutions can separate into two liquid phases. Either, or both, of the upper and lower critical solution temperatures may be obeyed, and many such examples exist (*3, 8*). For polymer solutions, the upper and lower type diagrams are usually separated one from the other. It is not necessary to focus on the unique closed-loop type, which is most common in dealing with phase separation of low molecular weight substances.

To analyze this problem more quantitatively, the free energy of mixing of a polymer solution is obtained by adding an enthalpy term of either the Van Laar or Scatchard form to Equation 12. Standard procedures are followed except for the substitution of v_1 for x_1. The critical conditions for phase separation, as well as the binoidal shape, can be calculated in the usual way from the free energy of mixing expression (*29*, Ch. XIII). Examples of the theoretical binoidal, as well as the experimentally determined curves for polystyrene fractions in cyclohexane for molecular weights ranging from 4.36×10^4 (PSD) to 1.27×10^5 (PSA), are shown in Figure 6 (*34*). The major conclusions predicted by theory are confirmed. They far outweigh the disparity that exists between theory and experiment. The most striking feature of Figure 6 is the dissymmetry in the phase diagram. The critical point occurs at a very low solute (polymer) concentration and is very dependent on molecular weight. Hence, upon phase separation, a very dilute phase is in equilibrium with a more concentrated phase. For binary systems of nonpolymeric liquids, the critical composition is almost invariably found in the middle of the diagram. This result has generally been accepted to be the norm. It is in fact of limited applicability. In actuality, a whole range in the character of these phase diagrams can be achieved. The physical reason for the striking dissymmetry of the phase diagram involving polymers results from the disparity in the size of the two components.

Liquid Crystals

Our discussion of polymer solutions so far has been limited to the more flexible chain types, which are highly coiled in solution. On the other extreme, we can also consider rigid rod-like molecules, which are geometrically highly asymmetric. This class of molecules is typified by the α-helical polypeptides and the aromatic polyamides and polyesters. The ratio of the length to the breadth of the molecule can be on the order of several hundred. Simple qualitative arguments can be presented to show that even

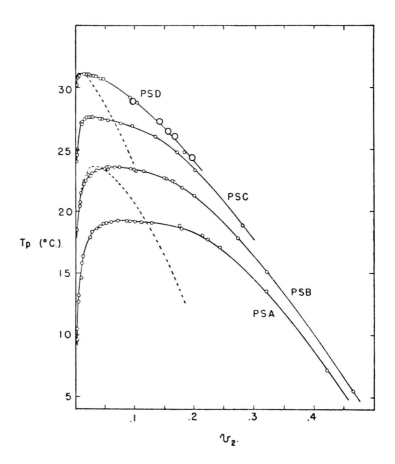

Figure 6. Precipitation temperature T_p for polystyrene fractions in cyclohexane plotted against the polymer volume fraction v_2. PSA $M_N = 4.36 \times 10^4$; PSB $M_N = 8.9 \times 10^4$; PSC $M_N = 2.5 \times 10^5$; PSD $M_N = 1.27 \times 10^6$. *Source:* Reference 34.

in the absence of any intermolecular interactions, an isotropic solution can only be maintained in the very dilute range. Because of the space-filling problem there is a concentration in the dilute range, depending on the molecular asymmetry, where the isotropy can no longer be maintained and a liquid-crystal nematic phase will form. This phenomenon, which is of great technological importance, can be given a straightforward explanation in terms of the geometric properties and space-filling requirements. A more quantitative description has been developed using lattice techniques in a natural extension of the Flory-Huggins theory (*35–37*).

Molecular properties

Studies of polymers can also give insight into some of the basic principles of statistical mechanics and thus molecular properties. One important factor that immediately emerges is the importance of averaging and the concept of average properties. We are almost always concerned with averages when dealing with macroscopic properties of low molecular weight substances. This point is brought home quite forcibly in studying polymers because of the tremendous configurational versatility of a long-chain molecule and the polydispersity in molecular weight that is invariably found in an assembly of such chains. For example, let us consider a polyethylene chain—$(CH_2)_n$—which contains n carbon atoms. If we allow, for the purpose of estimation, that the permissible rotational isomers are equally probable, then 3^n conformations are available to the molecule. For a chain of modest molecular weight, say $n = 10^4$, there are $10^{4,771}$ allowable conformations. A number of this size defies comprehension. Obviously, each of these distinctly different conformations cannot be enumerated and identified individually. One must therefore resort to the legitimate practice of averaging to describe the molecular structure and macroscopic properties.

The polydispersity in chain length that is usually found in natural and synthetic polymers makes necessary the use of average values in describing molecular weight. There are different methods of averaging that weigh different portions of the distribution and that are reflected in different types of measurement. For example, the number average molecular weight, M_N, is defined as

$$M_N = \frac{\sum_i M_i N_i}{\sum_i N_i} \qquad (15)$$

Here N_i is the number of molecules of species i having a molecular weight M_i. M_N is the molecular weight average determined by colligative type measurement. It is most sensitive to the lower molecular weights in the distribution. The weight average molecular weight M_w is defined as

$$M_w = \frac{\sum_i N_i M_i^2}{\sum_i N_i M_i} \qquad (16)$$

and can be experimentally determined by light scattering and equilibrium ultracentrifugation. It is most sensitive to the higher molecular weight species. The next higher average, M_z, is defined as

$$M_z = \frac{\sum_i N_i M_i^3}{\sum_i N_i M_i^2} \qquad (17)$$

which can also be determined by appropriate equilibrium centrifugation techniques. Still higher molecular weight averages can be defined following the scheme outlined, but they are not easily determined experimentally. It is important to note that each method of molecular weight determination yields its own characteristic molecular weight average.

A different type of averaging process is involved when studying the properties of a chain, as the following examples show. In a simple, but very useful, approximation, a long-chain molecule can be represented by a random walk in three dimensions with n steps each of length l. The probability $P(r)$ that the two ends of the chain are separated by the distance r is given by

$$P(r) = A r^2 \exp(-b^2 r^2) \qquad (18)$$

where the constant A is equal to $b^3/\pi^{3/2}$ and $b^2 = 3/2nl^2$. If r is replaced by v in Equation 18, then the Maxwell-Boltzmann distribution function for the velocity of a gas is obtained. The distribution function for the end-to-end distance can be used to obtain average properties in the usual manner. Instead of being applied to gases, different average sizes of a molecule can now be calculated. From Equation 18 the most probable value of r is zero. However, when only the magnitude of the end-to-end distance is considered, which is the quantity of physical interest,

$$\langle r^2 \rangle = l^2 n \qquad (19)$$

A linear dimension of the chain is directly proportional to the square root of the number of bonds. This is a very fundamental result. The analogy to the mean square displacement of a gas, and to Brownian motion, is clear, as is the reason that a real chain is highly coiled. It is an interesting exercise to compare these dimensions with those of a highly extended chain.

Many average properties, including dimensions, can be calculated for real chains having specified chemical repeating units. It is necessary to calculate the configurational partition function. We have been taught, and have traditionally assumed, that there are three contributions to the partition function: the translational, the vibrational, and the rotational. For long-chain molecules there is an additional contribution, that of the configurational partition function (38). The mathematical apparatus to calculate the configurational partition function is in place, and the calculation is exact (38). All that is needed is information with respect to the chemical nature of the repeating unit (or, in the case of copolymers, the sequence distribution of the co-units), the torsional potentials, and the interactions between nonbonded atoms. The latter two factors are the same as are found in the conformational analysis of nonpolymeric substances. Therefore, this aspect of the problem is identical. Although the calculation of this partition function can become quite involved, Mattice has shown how the basic concepts can be developed and actually used at the undergraduate level (5). Matrix methods are used to formulate the partition function. Students should be familiar with this technique from previous studies. However, as is illustrated in the example given below, the basic principles can be developed by using nothing more complicated than a 2 × 2 matrix.

Following Mattice (5), we take as an example the formulation of the configurational partition function of linear polyethylene. The well-established conformational analysis of the n-alkanes is followed, using the torsional potential for the internal C—C bond in n-butane. The rotational states are weighed in the same manner, and the steric interaction of successive gauche bonds of opposite sign—the pentane effect, to organic chemists—is also taken into account. If σ is the statistical weight of either gauche rotational state relative to the trans state, and ω is the extra weighing factor required for successive $g^{\pm}g^{\mp}$ placement, the statistical weight matrix can be written as

$$U = \begin{vmatrix} 1 & \sigma & \sigma \\ 1 & \sigma & \sigma\omega \\ 1 & \sigma\omega & \sigma \end{vmatrix} \quad (20)$$

The columns in the matrix index the state of the bond in question. The rows represent the preceding bond. The order of indexing is t, g^+, g^-. It can be shown that the configurational partition function can be expressed as (5)

$$Z = \text{row } (100) \, U^{n-2} \, \text{col } (111) \quad (21)$$

These expressions can be further simplified, in the case of polyethylene, by recognizing that g^+ and g^- have the same statistical weight as do g^+g^- and g^-g^+. Therefore, each configuration with one or more gauche placements has a twin with an identical statistical weight in which $g^{\pm}$ has been replaced by $g^{\mp}$. The configurational partition function can then be expressed as

$$Z = \text{row } (10) \, U^{n-2} \, \text{col } (11) \quad (22)$$

where

$$U = \begin{vmatrix} 1 & 2\sigma \\ 1 & \sigma(1+\omega) \end{vmatrix} \quad (23)$$

Thus, what originally appeared to be a completely unwieldy problem has been reduced to the manipulation of a 2 × 2 matrix.

When the number of bonds n in the chain is large, it is shown that (38)

$$Z = \text{const } \lambda_1^n$$

where λ_1 is the largest eigenvalue of the statistical weight matrix U. λ_1 is easily obtained from the 2 × 2 matrix of Equation 23. By utilizing these methods one can obtain average properties, such as the mean square end-to-end distance and the average dipole moment, for example, of real chains, no matter how complex they are. Other questions can also be asked and retrieved from the partition function. These include the trans-gauche ratio, the average sequence length of a particular rotational state, and others. The mathematical apparatus to solve this problem is exact. The only information that needs to be supplied is the torsional potentials and the nonbonded interactions. These factors are identical to those involved in nonpolymeric substances. Therefore, the analysis is not typical of polymers. Any problems that exist are common to all classes of substances.

The same techniques can be used to analyze order-disorder phenomena such as the popular one-dimensional helix-coil transition (5). Only the statistical weights for each state need to be supplied. A 2 × 2 matrix again results, and the analysis is very similar to the polyethylene problem. Thus, there is available an almost exact treatment of a complex and very interesting phenomenon that is suitable for presentation at the undergraduate level.

Rubber Elasticity

The understanding of the molecular basis of rubber elasticity represents a monumental accomplishment in relating the details of molecular structure to a macroscopic property for any type system. The subject is a classical casebook study that can be introduced into the undergraduate curriculum. Starting with the very early studies of Gough (*12*), which have been described earlier, subsequent thermodynamic analysis showed that for a rubber-like substance the major contribution to the retroactive force comes from the entropy decrease with deformation (*6; 29*, Ch. XI; *39; 40*). In contrast, for low molecular weight substances the major contribution to the retroactive force comes from energy contributions. In a formal sense, these differences account for the differences in elastic properties and the unique long-range elasticity of rubber-like substances. A representative set of force-length curves for the two types of substances is given in Figure 7. The simple solid, or low molecular weight substance, is relatively inelastic, as a large force is required to achieve a small extension. On the other hand, a five- to six-fold deformation is easily sustained by a rubber-like substance. Rubber-like elasticity is characterized by a system that can sustain large deformations and return to the original state when the deformation is removed. All long-chain molecules, irrespective of chemical type, display rubber-like deformations when in the disordered state above the glass temperature.

Figure 7. Schematic representation of force-extension curves for (a) rubber-like substance, (b) monomeric solid. *Source*: Reference 13.

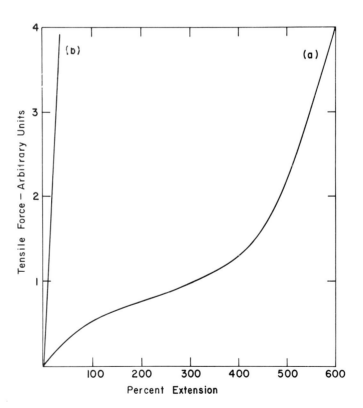

The unique elastic properties of a rubber-like substance can be put on a quantitative molecular basis. Although some very sophisticated theories, which are being continually improved, are now available, a more elementary approach gives results that are more than adequate for present purposes. We shall follow this approach here. The distribution of end-to-end vectors for a collection of chains in the undeformed state (which are very lightly cross-linked to prevent irreversible flow) is given by Equation 18. If we consider deformation in simple extension as an example, the vector components in the direction of elongation will be increased so that more extended configurations will be favored. The number of configurations available to the system will therefore decrease. Concomitantly there must also be a decrease in the entropy. This decrease in entropy is the source of the retroactive force. The retroactive force arises from a desire, or need, of the chains to return to the undeformed state, which is the most probable state having the maximum number of configurations and thus the maximum entropy. There is an obvious analogy in this phenomenon to the entropy decrease that occurs when the volume of an ideal gas is decreased.

The change in the number of configurations available to the system subject to a tensile force can be calculated from Equation 18 under the assumption of an affine deformation (*6; 29*, Ch. XI; *39; 40*). In this type of deformation, the change in the coordinates of the chain vectors is proportional to the macroscopic deformation. It is then found that in simple extension, the stress τ is related to the extension ratio α by the relation (*6; 29*, Ch. XI; *39*).

$$\tau = \frac{\nu}{V} RT\left(\alpha - \frac{1}{\alpha^2}\right) \tag{24}$$

where ν is the number of chains and V is the volume. Equation 24 can be considered to be a simplified, or ideal, equation of state for a rubber-like substance. According to this equation, at constant length, the stress should be directly proportional to the absolute temperature. This conclusion is in fact observed experimentally. There is analogy here to an ideal gas, where the pressure is proportional to the temperature at constant volume. The factor in front of the brackets in Equation 24 is the elastic modulus, which is directly predicted by molecular theory. Within the uncertainty in the chemical analysis for the number of crosslinks, and thus the number of chains, good agreement is obtained between the calculated and observed modulus.

The force-length relation predicted by Equation 24 is compared with typical experimental data in Figure 8. Very good agreement is found between theory and

Figure 8. Comparison between experiment and theory of the force-length relation of stretched natural rubber. Solid line is the theoretical expectation; open circles, the experimental points. *Source*: Reference 39.

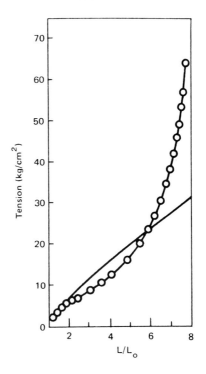

experiment at small extension ratios ($L/L_0 = \alpha$), and the agreement is good up to extension ratios of about 6. The very large deviations that are found at large extension ratios can be attributed to a phase change leading to an ordered state so that the molecular structural requirements of the theory are no longer satisfied.

The long-range elasticity of rubber-like substances can thus be put on a quantitative molecular basis. Throughout the analysis, the conventional laws and methods of physical chemistry have been used. This experience represents a classical achievement and demonstrates the unity in concept that exists in analyzing very diverse types of molecular substances.

Chemical kinetics

Space does not allow for a detailed development of the use of polymers or polymerization reactions in the general study of chemical kinetics. Although this subject has been discussed in detail elsewhere (7, 8, 41), some general observations would appear to be in order here. There are many examples that focus on the use of homogeneous and heterogeneous catalysis in polymerization reactions. Free-radical polymerization is most often catalyzed by soluble species that enter the process as initiators of the reaction (7). In contrast, ionic polymerization, involving insoluble Ziegler catalysts, i.e., certain organometallic coordination compounds, is heterogeneous in character and leads to stereoregular polymers (7, 42). In this situation not only is a chemical reaction catalyzed, but there is also control of the steric structure of the resulting chain.

There are two general categories of polymerization reactions. One is a stepwise or condensation polymerization, the other an addition or chain reaction. Condensation polymerization involves a series of step reactions that eventually leads to the formation of larger molecular weight species. In simple examples, the reaction of a diamine with a diacid leads to the formation of a polyamide or a diacid with a dialcohol to a polyester. (A very elegant experiment is described in the literature [43], which allows a nylon fiber to be formed in a simple classroom demonstration.) Each step in the reaction involves the elimination of a water molecule. The elementary reactions involve straightforward organic chemistry. However, the formation of high molecular weight polymers has some consequences that are very important to the general development and exposition of chemical kinetics. It has been experimentally demonstrated that the inherent reactivity of a functional group is independent of molecular weight (29, Ch. III and IV). Thus all the functional groups in the system have equal reactivity. This fact allows for the calculation of the molecular weight distribution as a function of the extent of the reaction (29, Ch. III and IV). When the species have higher functionality, branched molecules and three-dimensional network structures can be formed (29, Ch. III and IV).

Addition polymerization is representative of all chain reactions. In contrast to condensation polymerization, molecules are added to the growing chain without the elimination of any small species. The basic mechanism involves the activation of a monomer unit, usually by means of a catalyst; the propagation step, where monomer units are added to the growing chain; and the termination step. The process can also be complicated by chain transfer processes. By writing out the mechanism that is thought to apply to a specific polymerization, the chain length can be calculated (7; 29, Ch. III and IV; 41). Thus, by measuring the chain length as a function of conversion, the reaction can be analyzed and the details of the mechanism established.

Conclusion

I have tried to set forth a variety of examples involving polymeric substances. These illustrate some of the principles of physical chemistry that are applicable to all types of molecular systems and are suitable for inclusion in the first course. These examples expand the students' vistas, hold their interest, and simultaneously introduce them to a new set of molecules in a perfectly natural way. This can be accomplished without sacrificing any of the rigor and mental discipline that are necessary in the teaching of physical chemistry.

References

(1) Flory, P. J. *J. Chem. Educ.* **1973**, *50*, 732.
(2) Ferry, J. D. *J. Chem. Educ.* **1959**, *36*, 164.
(3) Krause, S. *J. Chem. Educ.* **1978**, *55*, 174.
(4) Mandelkern L. *J. Chem. Educ.* **1978**, *55*, 177.
(5) Mattice, W. L. *J. Chem. Educ.* **1981**, *58*, 911.
(6) Mark J. E. *J. Chem. Educ.* **1981**, *58*, 899.
(7) Morton, M. *J. Chem. Educ.* **1973**, *50*, 740.
(8) Core Committee, *J. Chem. Educ.* **1985**, *62*, 780, 1030.
(9) Mandelkern, L. *Crystallization of Polymers;* McGraw-Hill: New York, 1964.
(10) Pines, E.; Wun, K. L.; Prins, W. *J. Chem. Educ.* **1973**, *50*, 753.
(11) Wiegand, W. B. *Trans. Inst. Rubber Ind.* **1925**, *1*, 141.
(12) Gough, J. *Proc. Lit. and Phil. Soc.*, Manchester, 2nd Ser. *1*, 288 (1806).
(13) Mandelkern, L. *Introduction to Macromolecules*; Springer-Verlag: New York, 1983.
(14) Diorio, A. F.; Mandelkern, L.; Lippincott, E. R. *J. Phys. Chem.* **1962**, *66*, 2096.
(15) Matsuoka, S. *J. Polym. Sci.* **1960**, *42*, 511.
(16) McGear, P. L.; Duus H. C. *J. Chem. Phys.* **1952**, *20*, 1813.
(17) Flory, P. J. *J. Am. Chem. Soc.* **1956**, *78*, 5222.
(18) Flory, P. J. *Science* **1956**, *124*, 53.
(19) Oth, J.F.M.; Flory, P. J. *J. Am. Chem. Soc.* **1958**, *80*, 1297.
(20) Flory, P. J.; Spurr, O. K., Jr. *J. Am. Chem. Soc.* **1961**, *83*, 1308.
(21) Burt, C. T.; Mandelkern, L. *J. Mechanochem. Cell Motility* **1972**, *1*, 233.
(22) Mandelkern, L. *Ann. Rev. Phys. Chem.* **1964**, *15*, 421.
(23) Katchalsky, A.; Lifson, S.; Michaeli, I.; Zwick, M. In *Contractile Polymers*; Wasserman, A., Ed.; Pergamon Press: New York, 1960; p. 1.
(24) Steinberg, I. Z.; Oplatka, A.; Katchalsky, A. *Nature* **1966**, *210*, 568.
(25) Flory, P. J. *J. Chem. Phys.* **1949**, *17*, 223.
(26) Chiang, R.; Flory, P. J. *J. Am. Chem. Soc.* **1961**, *83*, 2857.
(27) Huggins, M. L. *J. Phys. Chem.* **1942**, *46*, 151; *Ann. N.Y. Acad. Sci.* **1942**, *41*, 1; *J. Am. Chem. Soc.* **1942**, *64*, 1712.
(28) Flory, P. J. *J. Chem. Phys.* **1942**, *10*, 51.
(29) Flory, P. J. *Principles of Polymer Chemistry*; Cornell University Press: Ithaca, N.Y., 1953.
(30) Flory, P. J. *Trans. Farad. Soc.* **1955**, *51*, 848.
(31) Kenney, J. F. *Polymer Eng. Sci.* **1968**, *8*, 216.
(32) Mandelkern, L. In *Physical Properties of Polymers*; Mark J. E., Ed.; American Chemical Society: Washington, D.C., 1984; p. 155.
(33) Krigbaum, W. R.; Flory, P. J. *J. Am. Chem. Soc.* **1953**, *75*, 5254.
(34) Shultz, A. R.; Flory, P. J. *J. Am. Chem. Soc.* **1952**, *74*, 4760.
(35) Flory, P. J. *Proc. Royal Soc.* **1956**, *234A*, 73.
(36) Flory, P. J. *Proc. Royal Soc.* **1956**, *234A*, 60.
(37) Flory, P. J. *Adv. Polym. Sci.* **1984**, *59*, 1.
(38) Flory, P. J. *Statistical Mechanics of Chain Molecules*; Interscience Publishers: New York, 1969.
(39) Treloar, L.R.G. *The Physics of Rubber Elasticity*; Clarendon Press: Oxford, 1949.
(40) Mark, J. E. In *Physical Properties of Polymers*; Mark, J. E., Ed.; American Chemical Society: Washington, D.C., 1984; p. 1.
(41) Mayo, F. R. *J. Chem. Educ.* **1959**, *36*, 157.
(42) Schultz, G. V. CHEMTECH **1973**, 200.
(43) Morgan, P.W.; Kwolek, S. L. *J. Chem. Educ.* **1959**, *36*, 182.

Chapter 8   APPLICATIONS OF THERMODYNAMICS AND KINETICS IN INORGANIC CHEMISTRY

EDWARD L. KING
Department of Chemistry
University of Colorado
Boulder, Colo. 80309

Many principles of physical chemistry are illustrated nicely by examples from the field of inorganic chemistry. There are advantages in employing real chemistry rather than A-plus-B-gives-C-plus-D chemistry in discussing thermodynamics and kinetics.

Thermodynamics and equilibrium

Thermodynamics of the Synthesis of Ammonia (1, 2)

For reactions of ideal gases, the change of enthalpy per mole of reaction does not depend upon the partial pressures of the reactant and product gases. The change of entropy per mole of reaction, on the other hand, does depend upon these partial pressures.

For an ideal gas, the molar entropy is a function of the pressure of the gas:

$$S = S^0 - R \ln (P/1 \text{ bar})$$

where S^0 is the molar entropy for the standard state, the hypothetical ideal gas with a pressure of 1 bar. This dependence, in fact, can be viewed as the root of the mass-action law. The ammonia-synthesis reaction nicely illustrates this point as well as others. Combination of the equations giving the entropy for each gaseous reactant and product yields for the synthesis of ammonia:

$$N_2(g) + 3 H_2(g) = 2 NH_3(g)$$

the equation for ΔS

$$\Delta S = \Delta S^0 - R \ln \frac{P_{NH_3}^2}{P_{N_2} \times P_{H_2}^3} \times 1 \text{ bar}^2.$$

At 400 K,

$$\Delta H = \Delta H^0 = -96{,}080 \text{ J mol}^{-1}$$

$$\Delta S = -210.66 \text{ J (K} \cdot \text{mol)}^{-1} - R \ln \frac{P_{NH_3}^2}{P_{N_2} \times P_{H_2}^3} \times 1 \text{ bar}^2$$

Combination of these equations with $\Delta G = \Delta H - T\Delta S$ yields for 400 K,

$$\Delta G = -11{,}820 \text{ J mol}^{-1} + RT \ln \frac{P_{NH_3}^2}{P_{N_2} \times P_{H_2}^3} \times 1 \text{ bar}^2$$

The numerical terms in the equations for ΔS and ΔG are, of course, the values of ΔS^0 [-210.66 J (K $\cdot$ mol)$^{-1}$] and ΔG^0 ($-11{,}820$ J mol^{-1}). With this value for ΔG^0, the equilibrium constant can be calculated:

$$K \times 1 \text{ bar}^2 = e^{-\Delta G^0/RT} = 35.0, \text{ or}$$

$$K = 35.0 \text{ bar}^{-2}$$

A state of chemical equilibrium exists at this temperature when the quotient of pressures

$$Q = \frac{P_{NH_3}^2}{P_{N_2} \times P_{H_2}^3}$$

has the value of the equilibrium constant: $K = 35.0$ bar^{-2}.

One can make conventional calculations using this equilibrium constant. Two are outlined in the box.

Equilibrium Calculations

$$N_2(g) + 3H_2(g) = 2NH_3(g) \qquad K = 35.0 \text{ bar}^{-2} \text{ at } 400 \text{ K}$$

Initial conditions

$$0.2500 \text{ mol } N_2 + 0.7500 \text{ mol } H_2$$
$$P = 1.000 \text{ bar} \quad V = 33.26 \text{ dm}^3$$

At equilibrium with $P = 1.000$ bar

$$P_{NH_3} = X$$

$$P_{N_2} = 0.2500 \text{ bar} - \frac{1}{4}X$$

$$P_{H_2} = 3P_{N_2} = 0.7500 \text{ bar} - \frac{3}{4}X$$

$$\frac{X^2}{27 \times \left(0.2500 \text{ bar} - \frac{X}{4}\right)^4} = 35.0 \text{ bar}^{-2}$$

yields $X = 0.493$ bar

$\therefore P_{N_2} = 0.127$ bar $\qquad P_{H_2} = 0.380$ bar

$$\frac{n_{NH_3}}{n_{N_2}} = \frac{n_{NH_3}}{0.250 \text{ mol} - \frac{n_{NH_3}}{2}} = \frac{X}{P_{N_2}} = \frac{0.493 \text{ bar}}{0.127 \text{ bar}} = 3.88_2$$

$\therefore n_{NH_3} = 0.330$ mol (66.0% yield)

$$V = \frac{1 \text{ mol} - 0.330 \text{ mol}}{1 \text{ mol}} \times 33.26 \text{ dm}^3 = 22.28 \text{ dm}^3$$

At equilibrium with $V = 33.26$ dm^3

$$P_{NH_3} = X$$

$$P_{N_2} = 0.2500 \text{ bar} - \frac{1}{2}X$$

$$P_{H_2} = 3P_{N_2} = 0.7500 \text{ bar} - \frac{3}{2}X$$

$$\frac{X^2}{27 \times \left(0.2500 \text{ bar} - \frac{X}{2}\right)^4} = 35.0 \text{ bar}^{-2}$$

yields $X = 0.302$ bar

$$n_{NH_3} = \frac{0.302 \text{ bar}}{1.000 \text{ bar}} \times 1 \text{ mol} = 0.302 \text{ mol} \quad (60.4\% \text{ yield})$$

$$P = 1.000 \text{ bar} \times \frac{1.000 \text{ mol} - 0.302 \text{ mol}}{1.000 \text{ mol}} = 0.698 \text{ bar}$$

Each calculation pertains to a system consisting initially of 0.2500 mol N_2 plus 0.7500 mol H_2 with a total pressure of 1.0000 bar and a volume of 33.26 dm^3. We note, in comparing the yields of ammonia in the two cases, that it is greater in the equilibrium state with the larger total pressure.

Now we will look at the approach to equilibrium in these two situations in terms of the Gibbs free energy for the constant-pressure case and in terms of the Helmholtz free energy for the constant-volume case. First the constant-pressure example will be considered. Equations for H/RT and S/R as a function of n, the amount of ammonia produced, are as follows (values for enthalpy and internal energy are based upon the usual convention, being relative to values for the elements in their standard states at the temperature in question):

$$H/RT = n \times \Delta H_f^0(NH_3)/RT$$

$$S/R = \left(0.2500 \text{ mol} - \frac{n}{2}\right)\left(\frac{S^0(N_2) + 3\,S^0(H_2)}{R}\right.$$

$$\left. - \ln\frac{0.2500 \text{ mol} - \frac{n}{2}}{1.0000 \text{ mol} - n} - 3\ln\frac{0.7500 \text{ mol} - \frac{3n}{2}}{1.0000 \text{ mol} - n}\right)$$

$$+ n\left(\frac{S^0(NH_3)}{R} - \ln\frac{n}{1.0000 \text{ mol} - n}\right)$$

Using the standard-state entropies: NH_3 203.59 JK^{-1} mol^{-1}, N_2 200.18 JK^{-1} mol^{-1}, and H_2 139.22 JK^{-1} mol^{-1}, one can calculate the basis for Figure 1a. We see a minimum in G/RT at $n = 0.330$ mol, the value calculated in the box.

Figure 1. (a) System of 0.2500 mol N_2 and 0.7500 mol H_2 at 400 K, $P = 1.000$ bar. H/RT, S/R, and G/RT as function of n_{NH_3}. Minimum at $n_{NH_3} = 0.330$ mol. (b) System of 0.2500 mol N_2 and 0.7500 mol H_2 at 400 K, $V = 33.26$ dm^3. E/RT, S/R, and A/RT as function of n_{NH_3}. Minimum at $n_{NH_3} = 0.302$ mol.

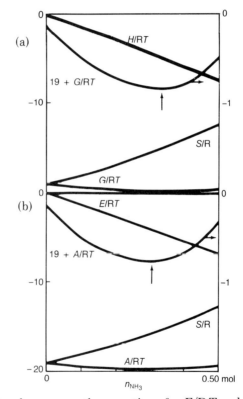

For the constant-volume case, the equations for E/RT and S/R are:

$$E/RT = n\left(\frac{\Delta H_f^0(NH_3)}{RT} + 1\right)$$

$$S/R = \left(0.2500 \text{ mol} - \frac{n}{2}\right)\left(\frac{S^0(N_2) + 3\,S^0(H_2)}{R}\right.$$

$$\left. - \ln\frac{0.2500 \text{ mol} - \frac{n}{2}}{1 \text{ mol}} - 3\ln\frac{0.7500 \text{ mol} - \frac{3n}{2}}{1 \text{ mol}}\right)$$

$$+ n\left(\frac{S^0(NH_3)}{R} - \ln\frac{n}{1 \text{ mol}}\right)$$

In Figure 1b, the minimum in A/RT occurs at $n = 0.302$ mol, the value calculated in the box. The minimum in each plot occurs at the degree of advancement in the reaction at which $Q = K$.

Figure 2. The reaction $N_2(g) + 3 H_2(g) = 2 NH_3(g)$ at 400 K (where $K = 35.0$ bar^{-2}) (a) $\Delta H/RT$, $\Delta S/R$, and $\Delta G/RT$ as function of $\ln (Q \times 1 \text{bar}^2)$ (b) $\Delta S_{\text{system}}/R$, $\Delta S_{\text{surrounding}}/R$, and $\Delta S_{\text{universe}}/R$ as function of $\ln (Q \times 1 \text{bar}^2)$.

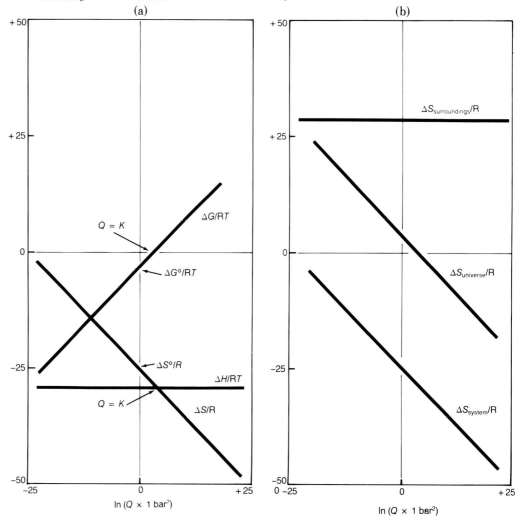

Discussion of graphs such as Figures 1a and 1b may entail the concept of entropy of mixing (3), but such emphasis is unnecessary. The equation for molar entropy $S = S^0 - R \ln (P/1 \text{ bar})$ is the entire story. This point has been made recently (3a). Clearly, entropy of mixing plays no role in establishing the position of equilibrium in the change of state

$$H_2O(l) = H_2O(g)$$

occurring in a vessel having a particular volume less than nRT/vp, where vp is the vapor pressure of water at the temperature in question. Calculations of E/RT, S/R, and A/RT for this process yield a minimum in A for vaporization of that amount of water corresponding to the vapor pressure.

The position of equilibrium can be discussed in other ways. Equilibrium exists for any set of partial pressures that makes $T\Delta S$ for such conditions equal to ΔH, which we take to be independent of the conditions and equal to ΔH^0; it is under these conditions that $\Delta G = 0$. This is illustrated in Figure 2a (4), where a plot of $\Delta H/RT$, $\Delta S/R$, and $\Delta G/RT$ versus $\ln (Q \times 1 \text{ bar}^2)$. The line giving $\Delta G/RT$ is the difference between the lines giving $\Delta H/RT$ and $\Delta S/R$. Indicated on the graph is the place where $Q = K$; this is where $\Delta S/R = \Delta H/RT$. The values of $\Delta G/RT$ and $\Delta S/R$ where $Q = 1$ bar^{-2} also are indicated. The intersections of the plots of $\Delta G/RT$ and $\Delta S/R$ with the $Q = 1$ bar^{-2} axis occur where $\Delta S = \Delta S^0$ and $\Delta G = \Delta G^0$.

The direction in which reaction occurs to approach equilibrium depends on the value of Q relative to the value of K (35.0 bar^{-2} for the situation at hand). If $Q > K$, reaction does not occur as written. Rather it occurs from right to left. If $Q < K$, reaction occurs as written. If $Q = K$, the system is at equilibrium.

Related to Figure 2a is Figure 2b, in which $\Delta S_{system}/R$, $\Delta S_{surroundings}/R$, and $\Delta S_{universe}/R$ are plotted versus $\ln (Q \times 1 \text{ bar}^2)$. The line $\Delta S_{system}/R$ is the line $\Delta S/R$ in Figure 2a; the line $\Delta S_{surroundings}/R$ is, with the sign changed, the line $\Delta H/RT$ in Figure 2a; and the line $\Delta S_{universe}/R$ is, with the sign changed, the line $\Delta G/RT$ from Figure 2a. Thus we can discuss the approach to equilibrium in terms of the balance between ΔS_{system} and $\Delta S_{surroundings}$. If $Q < K$, $\Delta S_{surroundings}$, a positive quantity, exceeds in magnitude ΔS_{system}, a negative quantity, thereby making occurrence of reaction as written associated with an increase of entropy of the universe. For $Q > K$, $\Delta S_{surroundings}$ does not exceed in magnitude ΔS_{system}; their sum is negative. To approach equilibrium, reaction occurs in the reverse direction. The emphasis in Figure 2b is that stressed by Atkins (5). Consideration together of the approaches in the box and Figures 1 and 2 simply emphasizes the utility of the conventional calculation involving the equilibrium constant, outlined in the box.

The Thermal Decomposition of Water by a Closed-Cycle Thermochemical Process (6)
Thermodynamically infeasible reactions can be made to occur as the net result of an appropriate sequence of thermodynamically feasible reactions. The development of this counterintuitive assertion applied to the thermal decomposition of water illustrates further the interplay of changes of enthalpy and entropy and the role of temperature in determining the position of equilibrium in a chemical reaction.

Equilibrium in the thermal decomposition of water vapor

$$H_2O(g) = H_2(g) + \frac{1}{2} O_2(g)$$

is unfavorable except at very high temperature; the values of $K/\text{bar}^{1/2}$ as a function of temperature are: 298.2 K, 9.0×10^{-41}; 1000 K, 8.7×10^{-11}; 2000 K, 2.9×10^{-4}; 3000 K, 0.045. The value of this equilibrium constant is 1.00 bar$^{1/2}$ at ~4300 K (*1*). (At temperatures at which the decomposition of water vapor occurs to an appreciable extent, there also are significant amounts of species in addition to $H_2O(g)$, $H_2(g)$, and $O_2(g)$. At 3000 K, with the total pressure equal to 1.000 bar, approximate partial pressures (P/bar) for all of the species are: H_2O, 0.65; H_2, 0.14; O_2, 0.045; OH, 0.091; H, 0.058; and O, 0.024 [*1*].) Current interest in the thermal decomposition of water is not focused on the unassisted process at practically inaccessible temperatures, but rather is directed at closed cycles of reactions (Solvay clusters [*7, 8*])[1], each step of which occurs at lower temperatures; these Solvay clusters have water decomposition as their net result. The simplest imaginable closed cycles involve two steps:

Oxide Cycle

$$X + H_2O = XO + H_2$$

$$XO = X + \frac{1}{2} O_2$$

Hydride Cycle

$$Y + H_2O = H_2Y + \frac{1}{2} O_2$$

$$H_2Y = H_2 + Y$$

The intermediate chemicals X and XO or Y and H$_2$Y are not consumed; rather they are recycled. In this respect, each sequence of reactions resembles a common mechanism for catalysis, but this resemblance is misleading. The decomposition of water at the lower temperatures in question (<2000 K)[2] is thermodynamically unfavorable, and no catalyst operating at a *particular temperature* can increase the yield of hydrogen and oxygen above the equilibrium values. The substances X, XO, Y, and H$_2$Y are not catalysts for the decomposition of water.

---

[1] May and Rudd (*8*) use the term "Solvay cluster" for a series of reactions involving "intermediate chemicals and chemical reactions to bypass an important but unwilling chemical reaction." The Solvay soda ash process for the conversion of sodium chloride plus calcium carbonate into sodium carbonate plus calcium chloride (with $\Delta G^0 = +99$ kJ mol^{-1} at 298.2 K) is, like the decomposition of water, a thermodynamically infeasible reaction.

[2] Even 2000 K is too high a temperature to be achieved in nuclear reactors, a heat source proposed for use in thermal decomposition of water. Wentorf and Hanneman (*9*) suggest 1200 K as an upper limit achievable in the near future.

Figure 3. ΔH^0 versus ΔS_e for the reaction

$$H_2O(g) = H_2(g) + \frac{1}{2}O_2(g)$$

The plot is based upon $\Delta H^0 = 247.89$ kJ mol^{-1}, and $\Delta S^0 = 55.31$ JK^{-1} mol^{-1}, values appropriate for 1000 K. The slopes of the lines $T = \Delta H^0/\Delta S_e$ for K/bar$^{1/2} = 1.00, 1.00 \times 10^{-4}, 1.00 \times 10^{-8}$, and 1.00×10^{-12} are 4490 K, 1880 K, 1190 K, and 870 K, respectively. The correct temperatures for these values of K are 4300 K, 1870 K, 1190 K, and 870 K, respectively (1).

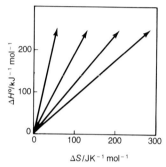

Figure 4. ΔH^0 versus ΔS_e for the reaction

$$2Ag(s) + H_2O(g) = Ag_2O(s) + H_2(g) \quad I$$
$$Ag_2O(s) = 2Ag(s) + \frac{1}{2}O_2(g) \quad II$$

with partial pressures of H$_2$O, H$_2$, and O$_2$ such that $K_I = 1.00 \times 10^{-4}$ and $K_{II} = 1.00$ bar$^{1/2}$. (For the construction of this figure, the values of ΔH^0 and ΔS^0 are those for 298.2 K.)

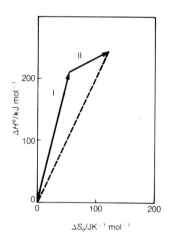

To discuss this further, we will focus attention on several oxide cycles and on the temperature at which the equilibrium constant for a reaction has a particular value. Rearrangement of the equation

$$-RT \ln K = \Delta H^0 - T\Delta S^0$$

yields

$$T = \frac{\Delta H^0}{\Delta S^0 - R \ln K}$$

for the temperature at which the equilibrium constant has a prescribed value. The denominator of this equation is simply the value of ΔS for the set of partial pressures that correspond to equilibrium at the temperature in question; this quantity will be designated as ΔS_e. If we assume that $\Delta C_p^0 \cong 0$, the values of ΔH^0 and ΔS^0 are independent of temperature. With a particular set of values of ΔH^0 and ΔS^0 selected for the range of temperature under consideration, the temperature satisfying the equation just given is the slope of a vector from the origin to the point $\Delta H^0, \Delta S_e$. Figure 3 shows plots for four values of the equilibrium constant for the reaction that is the focus of our attention.

Each reaction that is part of a sequence of reactions whose sum is the decomposition of water can be represented as a vector in the $\Delta H^0, \Delta S_e$ plane. The sum of these vectors is the vector for the water decomposition reaction. This will be illustrated first for an oxide cycle involving an easily decomposed oxide, Ag$_2$O:

$$2Ag(s) + H_2O(g) = Ag_2O(s) + H_2(g) \quad I$$
$$Ag_2O(s) = 2Ag(s) + \frac{1}{2}O_2(g) \quad II$$

Each of these reactions is endothermic (at 298.2 K, $\Delta H_I = 211.3$ kJ mol^{-1}, $\Delta H_{II} = 30.5$ kJ mol^{-1}) and for reaction I, $\Delta S_I^0 = -21.8$ JK^{-1} mol^{-1}. The vector for reaction I has a positive slope only for values of $K < 0.073$, for which $\Delta S_e > 0$ ($\Delta S_e = -21.8$ JK^{-1} mol^{-1} $-$ 8.315 JK^{-1} mol^{-1} $\times \ln(0.073) = 0$). Figure 4 displays the vectors for $K_I = 1.00 \times 10^{-4}$ ($\Delta S_e = \Delta S_I^0 + 76.58$ JK^{-1} mol^{-1}) and $K_{II} = 1.00$ bar$^{1/2}$. The sum of these vectors is the vector that corresponds to the water decomposition reaction having an equilibrium constant $K = K_I \times K_{II} = (1.00 \times 10^{-4}) \times (1.00$ bar$^{1/2}) = 1.00 \times 10^{-4}$ bar$^{1/2}$ at a temperature that is the slope of the resultant vector; this temperature is the reciprocal of the weighted average of the reciprocals of the temperatures at which reactions I and II have the assigned values of the equilibrium constants, the weighting factors being the values of ΔH^0 for the reactions. That is, at a temperature given by

$$\frac{1}{T} = \frac{1}{T_1} \times \frac{\Delta H_I^0}{\Delta H_I^0 + \Delta H_{II}^0} + \frac{1}{T_2} \times \frac{\Delta H_{II}^0}{\Delta H_I^0 + \Delta H_{II}^0}$$

the equation $K(T) = K_I(T_1) \times K_{II}(T_2)$ holds, as it does for $T = T_1 = T_2$. But this figure shows that reaction I of this cycle has a favorable equilibrium only at temperatures even higher than the temperature at which the unassisted decomposition of water occurs. This example involving *two endothermic reactions* illustrates an important point revealed clearly in the graph: *No series of reactions, each with $\Delta H^0 > 0$ and $\Delta S_e > 0$, can accomplish the decomposition of water at a temperature below the temperature at which the unassisted decomposition would occur.* The slope of at least one of the vectors added must be more positive than the slope of the resultant vector, which represents the temperature at which the unassisted reaction occurs.

Since no closed cycle consisting only of endothermic reactions is workable for the objective under consideration, let us consider a two-reaction cycle involving an exothermic step. An oxide cycle involving a difficultly decomposed oxide, K$_2$O,[1] is such a cycle:

$$2K(l) + H_2O(g) = K_2O(s) + H_2(g) \quad I$$
$$K_2O(s) = 2K(l) + \frac{1}{2}O_2(g) \quad II$$

---

[1] This cycle involving liquid potassium must be viewed as hypothetical because the normal boiling point of potassium is 1043.7 K, a temperature below that at which the endothermic step of the cycle could occur. If a cycle involving monoatomic gaseous potassium were considered, each temperature at which $K = 1.00$ would be lower.

The first of these reactions is exothermic; the second is endothermic (at 1000 K, $\Delta H^0_I = -104.0$ kJ mol^{-1}, $\Delta H^0_{II} = +351.8$ kJ mol^{-1}); and the standard entropy changes (-72.0 JK^{-1} mol^{-1} and 127.5 JK^{-1} mol^{-1}, respectively) are such that each reaction has a favorable equilibrium at a temperature below that for which the water decomposition equilibrium is favorable

$$K_I = 1.00 \text{ at } T = \frac{-104.0 \times 10^3 \text{ J mol}^{-1}}{-72.0 \text{ JK}^{-1} \text{ mol}^{-1}} = 1440 \text{ K}$$

$$K_I > 1.00 \text{ at } T < 1440 \text{ K}$$

and

$$K_{II} = 1.00 \text{ bar}^{1/2} \text{ at } T = \frac{+351.8 \times 10^3 \text{ J mol}^{-1}}{+127.5 \text{ JK}^{-1} \text{ mol}^{-1}} = 2760 \text{ K}$$

$$K_{II} > 1.00 \text{ at } T > 2760 \text{ K}$$

Figure 5, the plot of ΔH^0 versus ΔS^0 for this cycle illustrates a second important point: *A closed cycle of reactions with at least one exothermic reaction may accomplish the decomposition of water below the temperature at which the unassisted reaction occurs to the same extent.*

Figure 5. ΔH^0 versus ΔS^0 for the reactions
2K(l) + H$_2$O(g) = K$_2$O(s) + H$_2$(g)   I
K$_2$O(s) = 2K(l) + $\frac{1}{2}$ O$_2$(g)   II
Values of ΔH^0 and ΔS^0 are those for 1000 K (*1*).

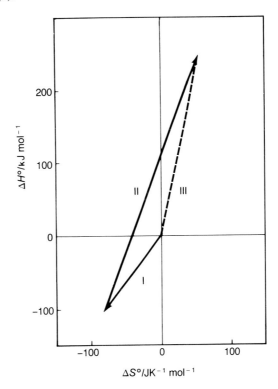

The potassium oxide cycle in which an endothermic reaction occurs at a high temperature and an exothermic reaction occurs at a low temperature is an example of a thermochemical engine (*7*), in which some of the energy flowing as heat into the system at the high temperature is converted into chemical energy, the energy of hydrogen and oxygen relative to that of water; the remainder is expelled as heat at the low temperature. Although this example illustrates the point being discussed, the temperature required for decomposition of potassium oxide is impractically high. This raises the question of whether a two-reaction oxide cycle operating between practical temperatures is possible.

The information conveyed in Figure 5 can equally well be conveyed if the origin for the vector for reaction II, the oxide decomposition reaction, is at the origin of the coordinate system. Figure 6 is such a plot,

Figure 6. ΔH^0 versus ΔS^0 for the reaction
$\frac{1}{m} X_n O_m(s) = \frac{n}{m} X(s) + \frac{1}{2} O_2(g)$ at 298.2 K.
The points are for the compounds (1) CaO, (2) Li$_2$O, (3) $\frac{1}{3}$ Al$_2$O$_3$, (4) TiO, (5) ZnO, (6) CO$_2$(g) (the product is CO(g)), (7) Na$_2$SO$_4$ (the product is Na$_2$SO$_3$). The vertical line $\Delta S^0 = 102.5$ JK^{-1} mol^{-1} is at $\Delta S^0 = 1/2\, S^0(O_2(g))$. The dashed line is the vector corresponding to the water-decomposition reaction. The lines radiating from the origin correspond to $T_2 = \Delta H_2^0/\Delta S_2^0 = 1300$ K and 1200 K. The lines radiating from the point $\Delta H^0 = 241.8$ kJ mol^{-1}, $\Delta S^0 = 44.4$ JK^{-1} mol^{-1} correspond to $T_1 = \Delta H_1^0/\Delta S_1^0 = 500$ K and 400 K. Thus, if a two-reaction oxide cycle were to have $K_1 = 1$ between 400 K and 500 K and $K_2 = 1$ bar$^{1/2}$ between 1200 K and 1300 K, the point for the oxide would lie in the shaded region.

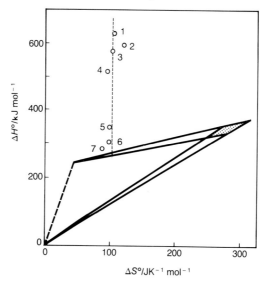

showing values of ΔH^0 and ΔS^0 for the decomposition of a number of oxides X$_n$O$_m$ to give 1/2 mol of O$_2$. The value of ΔS^0 for each of these reactions is close to $1/2 S^0(O_2(g))$ (102.5 JK^{-1} mol^{-1} at 298.2 K) because the quantity

$$\left\{ \frac{n}{m} S^0(X) - \frac{1}{m} S^0(X_n O_m) \right\}$$

is close to zero (0 ± 20 JK^{-1} mol^{-1}) if both the oxide X$_n$O$_m$ and its reduced form X are in a condensed state (or both are gases) (10, 11). (The decomposition of gaseous water deviates greatly from this generalization because hydrogen has a very small molar entropy.) This figure also shows vectors with slopes corresponding to practical values of temperature, both the low temperature at which the exothermic reaction occurs, $T_1 = 400$ K or 500 K, and the high temperature at which the endothermic reaction occurs, $T_2 = 1200$ K or 1300 K. The shaded region of this graph defined by the conditions $K_I = 1.00$ at $T = 400$ K to 500 K and $K_{II} = 1.00$ bar$^{1/2}$ at $T = 1200$ K to 1300 K is far from the coordinates for known oxides. Clearly, no known oxide will be the basis for a practical two-reaction oxide cycle, a point made in the literature cited in Reference 6. (Quantitative features of the graph would be altered if smaller values of the equilibrium constants were employed or if the experimental points were appropriate for another temperature, but the conclusion regarding impracticality of a two-reaction oxide cycle would not be changed.)

It is possible, however, to meet the critical temperature requirements for the individual reactions with a sequence having more than two reactions. A two-reaction cycle involving the water-gas reaction, like the silver oxide cycle, involves two endothermic steps, and it cannot advantageously accomplish the decomposition of water. But the water-gas reaction can be made part of a three-reaction closed cycle

$\qquad$ C(s) + H$_2$O(g) = CO(g) + H$_2$(g) $\qquad\qquad\qquad\qquad\qquad$ I

$\qquad$ CO(g) + 2Fe$_3$O$_4$(s) = C(s) + 3Fe$_2$O$_3$(s) $\qquad\qquad\qquad\qquad$ II

$$3Fe_2O_3(s) = 2Fe_3O_4(s) + \frac{1}{2}O_2(g) \qquad \text{III}$$

the individual steps of which do occur at lower temperatures than the temperature at which the unassisted reaction occurs. The ΔH^0 versus ΔS^0 graph for this closed cycle, a process studied by Marchetti and de Beni (see Chao [12]), is given in Figure 7.

Figure 7. The three-step closed cycle
$C(s) + H_2O(g) = CO(g) + H_2(g)$ I
$CO(g) + 2Fe_3O_4(s) = C(s) + 3Fe_2O_3(s)$ II
$3Fe_2O_3(s) = 2Fe_3O_4(s) + 1/2\ O_2(g)$ III

The values of ΔH^0 and ΔS^0 used in constructing this graph are those appropriate for 1000 K: $\Delta H_I^0 = 135.88$ kJ mol^{-1}, $\Delta S_I^0 = 143.60$ JK^{-1} mol^{-1}; $\Delta H_{II}^0 = -127.69$ kJ mol^{-1}, $\Delta S_{II}^0 = -227.37$ JK^{-1} mol^{-1}; and $\Delta H_{III}^0 = 239.70$ kJ mol^{-1} and $\Delta S_{III}^0 = 139.30$ J K^{-1} mol^{-1} (1).

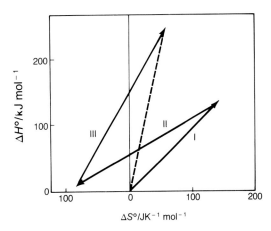

If values of ΔH^0 and ΔS^0 for 1000 K are used in the calculations, the temperature limits for the reactions are: (I) $K > 1$ bar at $T > 950$ K; (II) $K > 1$ bar^{-1} at $T < 560$ K; and (III) $K > 1$ bar$^{1/2}$ at $T > 1720$ K. These calculated temperature limits are close to those mentioned in a presentation of this process (12), 970 K, 520 K, and 1670 K, respectively.

Many Solvay clusters that involve more than three reactions have been proposed for the decomposition of water, and some of these involve reactions occurring at temperatures where water is a liquid. No new concepts would be introduced by consideration of such reaction sequences. The discussion presented here has focused on diagrams that reveal the temperature at which a reaction has a favorable equilibrium. Provision is made in these diagrams for altering the criterion for a favorable equilibrium by adjusting the value of K in the calculation of ΔS_e ($\Delta S_e = \Delta S^0 - R \ln K$), which is the abscissa in the diagram. Other types of graphs have been employed to focus attention upon other aspects of this subject, e.g., entropy versus temperature (13), enthalpy versus temperature (12), and free energy versus temperature (8).

The Correlation of Equilibrium Constants for Successive Steps of Metal-Ion Ligation

Constant Coordination-Number Systems

Most metal ions with charge 2+ or greater exist in aqueous solution as discrete hydrated species. For many, a particular coordination number and geometry are dominant, e.g., tetrahedral $Be(OH_2)_4^{2+}$ and octahedral $Cr(OH_2)_6^{3+}$. In the simplest mode of ligation, a monodentate ligand X replaces a coordinated water molecule in each of a series of stepwise reactions, as in the beryllium(II)-fluoride system. A careful study of this system has been made by Mesmer and Baes (14) using fluoride-ion-electrode and hydrogen-electrode measurements. The primary quantity derived from the data is $\bar{n}$, the average number of fluoride ion ligands bound to beryllium(II), as a function of the concentration of fluoride ion. Figure 8 presents $\bar{n}$ versus $\log[F^-]$ for 25.0 °C in 1 molal $NaClO_4$.

Figure 8. Beryllium(II)–fluoride complexes. $\bar{n}$ as a function of log[F$^-$]. Representative points from Reference 14. Solid line calculated as described in text. Dashed line is calculated $\bar{n} = 4\kappa[\text{F}^-]/(1 + \kappa[\text{F}^-])$.

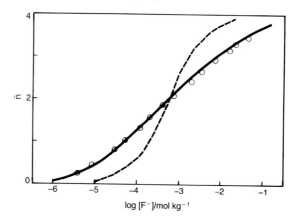

Also shown in this figure are calculated values of $\bar{n}$ versus log[F$^-$] for complexing in a tetrahedral $M(OH_2)_{4-n}F_n^{m-n}$ system governed by equilibrium constants related to one another by the statistical factor. We see that ligation of beryllium(II) ion by fluoride ion extends over a larger range of ligand concentration than corresponds to the statistical case. For the beryllium(II)-fluoride system, the change of $\bar{n}$ from 1 to 3 occurs over an approximately 200-fold change of concentration of fluoride ion. For the statistical case, this change of ligation occurs over a 9-fold change of ligand concentration. This is an example of anticooperativity; the binding of each successive ligand occurs with a smaller affinity. (Later we will discuss cooperative ligation.) Before discussing the beryllium(II)-fluoride system, we will consider the statistical contribution to equilibrium constants for reactions of multisite reagents.

The equilibrium constant for formation of the complex $M(OH_2)_{4-n}F_n^{m-n}$ from its precursor $M(OH_2)_{5-n}F_{n-1}^{m+1-n}$ contains a statistical factor that can be viewed in a number of ways:

1. The statistical factor is the ratio of the number of sites available for the ligand to add to the reactant divided by the number of sites occupied by ligands in the product of the reaction, ligands that can dissociate in the reverse reaction, i.e., $(5 - n)/n$. For the four stepwise reactions under consideration, the statistical factors are

n	1	2	3	4
$(5 - n)/n$	4/1	3/2	2/3	1/4

2. The statistical factor is the ratio of the symmetry numbers, σ, of the reactant metal species and product metal species (15). (The symmetry numbers of reactant ligand and water are common to each reaction and are omitted in the present discussion.) Thus, for the stepwise reactions under consideration, we have these ratios:

n	1	2	3	4
$\dfrac{\sigma(M(OH_2)_{5-n}F_{n-1})}{\sigma(M(OH_2)_{4-n}F_n)}$	12/3	3/2	2/3	3/12

3. The statistical factor for the overall reaction forming $M(OH_2)_{4-n}F_n^{m-n}$ from $M(OH_2)_4^{m+}$ is the number of ways in which the n coordinated ligands and the $4 - n$ coordinated water molecules can be arranged in the product species,

$$_4C_n = \frac{4!}{n!\,(4-n)!}$$

n	1	2	3	4
$_4C_n$	4	6	4	1

These statistical factors are the products of the statistical factors for the stepwise reactions

$$4 \qquad 4 \times \frac{3}{2} = 6 \qquad 4 \times \frac{3}{2} \times \frac{2}{3} = 4 \qquad 4 \times \frac{3}{2} \times \frac{2}{3} \times \frac{1}{4} = 1$$

These values also are the number of the geometric features of the tetrahedron defined by the n ligands in $M(OH_2)_{4-n}F_n^{m-n}$:

1 F   defines a corner of which there are 4
2 Fs   define an edge of which there are 6
3 Fs   define a face of which there are 4
4 Fs   define the tetrahedron of which there is 1

The dependence of $\bar{n}$ upon ligand concentration for a system with N statistically related equilibrium constants is simply (16)

$$\bar{n} = N \times \left(\frac{\kappa[X]}{1 + \kappa[X]}\right)$$

in which the factor $\kappa[X]/(1 + \kappa[X])$ is the equation for $\bar{n}$ for a one-site system. (If the N sites are truly independent, it is immaterial whether they are on one molecule or N molecules.)

The four stepwise equilibrium constants

$$K_n = \frac{[Be\ F_n^{2-n}]}{[Be\ F_{n-1}^{3-n}][F^-]} \times 1\ mol\ kg^{-1}$$

$K_1 = 7.9 \times 10^4$ $K_2 = 5.8 \times 10^3$
$K_3 = 6.1 \times 10^2$ $K_4 = 27$

evaluated from the data in Figure 8 have relative values: $1 : 0.073 : 0.0077 : 0.00034$, not the statistically expected $1 : 0.375 : 0.167 : 0.0625$.

A model used by Pauling (17) to correlate equilibrium constants for oxygen binding by hemoglobin, a cooperative system, may be applied to this anticooperative system. In this model, the cause for deviation from statistical behavior is ligand–ligand interaction. In this system, there are three different ligand–ligand interactions, water–water (W-W), water–fluoride (W-F), and fluoride–fluoride (F-F). The number of each type of ligand–ligand interaction in the beryllium (II)–fluoride complexes is given in Table I.

Table I. Number of Ligand-Ligand Interactions

Species	W-W	W-F	F-F
$Be(OH_2)_4^{2+}$	6	—	—
$Be(OH_2)_3F^+$	3	3	—
$Be(OH_2)_2F_2$	1	4	1
$Be(OH_2)F_3^-$	—	3	3
BeF_4^{2-}	—	—	6

Note: The total of six ligand–ligand interactions corresponds to the six edges of a tetrahedron.

This table provides the basis for expressing the four stepwise equilibrium constants in terms of an intrinsic constant, κ, for the ligand replacement and factors represented by lower-case letters for the changes of ligand–ligand interactions in the reaction. The resulting equations involve two parameters, q and α_t:

$$K_1 = 4q \qquad K_2 = \frac{3}{2} q \alpha_t$$

$$K_3 = \frac{2}{3} q \alpha_t^2 \qquad K_4 = \frac{1}{4} q \alpha_t^3$$

where $q = \kappa \times (wf/ww)^3$ and $\alpha_t = (ff)(ww)/(wf)^2$. That is, q is the intrinsic equilibrium constant for replacing a coordinated water molecule with a fluoride ion and concomitantly replacing three water–water interactions with three water–fluoride interactions. And α_t in an equilibrium constant is the factor attributable to replacing in a tetrahedral complex two water–fluoride interactions with one each water–water and fluoride–fluoride interactions. It should be noted that the exponent to which α_t is raised in $\beta_n (\beta_n = \pi K_1 \ldots K_n)$ is $n(n-1)/2$, the number of pairs of fluoride ligands in $Be(OH_2)_{4-n}F_n^{2-n}$. This model corresponds to a constant ratio of successive statistically corrected equilibrium constants (14):

$$\frac{\frac{2}{3}K_2}{\frac{1}{4}K_1} = \frac{\frac{3}{2}K_3}{\frac{2}{3}K_2} = \frac{4K_4}{\frac{3}{2}K_3} = \alpha_t$$

This model is reasonably successful in correlating the beryllium(II)-fluoride ligation data. The parameters $q = 2.0 \times 10^4$, $\alpha_t = 0.19$ yield calculated values of the equilibrium constants ($K_n \times 1 \text{ mol kg}^{-1}$):

$K_1 = 8.0 \times 10^4$ $K_2 = 5.7 \times 10^3$
$K_3 = 4.8 \times 10^2$ $K_4 = 34$

and the solid line in Figure 9 is calculated using these parameters. A value of α_t less than 1 corresponds to anticooperativity rooted in a destabilizing effect of like-ligand interactions. The electrostatic repulsion of the anionic fluoride ligands must be the dominant factor in making $\alpha_t = 0.19$.

Figure 9. Zinc(II) complexes with ammonia and cyanide ions. (Data for ammonia system from Reference 21; data for cyanide system from Reference 22 [representative points]). Solid lines calculated as described in text. Dashed lines: $\bar{n} = 4\kappa[X]/(1 + \kappa[X])$.

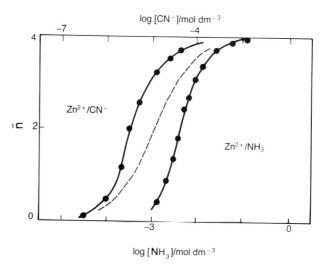

Data for other anticooperative systems of complexes with assumed octahedral coordination are correlated adequately by a model analogous to this (18). For octahedral systems $M(OH_2)_{6-n}X_n$, $q = \kappa(wx/ww)^4$, $\alpha_o = (xx)(ww)/(wx)^2$. For scandium(III)-fluoride complexes (19), $\alpha_o = 0.099$; for chromium(III)-methanol complexes, $\alpha_o = 0.615$ (18); and for chromium(III)-ethanol complexes, $\alpha_o = 0.545$ (18). For each of the latter two systems with a neutral ligand replacing a coordinated water molecule, the greater steric requirement of the alcohol-molecule ligands probably is the root of the anticooperativity. Octahedral systems are more complex because of the possibility of isomeric bis, tris, and tetra species. This ligand-ligand interaction model corresponds to the ratio of isomeric species, after correction for the statistical factor, being α_o.

Variable Coordination-Number Systems and Cooperativity (20)

There are systems of metal-ion complexes displaying cooperativity that are not correlated with the model just outlined. (Values of α greater than 1 correspond to cooperativity.) For some of these systems, the dominant aquametal ion and the fully ligated complex have different coordination numbers. Complexes of zinc(II) with ammonia (21) and with cyanide ion (22) are such systems. Hexaaquazinc(II) ion is dominant in acidic aqueous solutions containing no complexing agent (23), but tetraamminezinc(II) ion and tetracyanozinc(II) ion, both assumed to be tetrahedral species, are the dominant complexes at high concentrations of ligand. (There is some evidence for tetraaquazinc(II) as a minor species in aqueous solution [23].) Figure 9 shows $\bar{n}$ versus the logarithm of the ligand concentration for these systems.

Also shown is the curve for statistical binding of four ligands. We see that ligation for these systems occurs over a much smaller change of ligand concentration than corresponds to the statistical model. For the zinc(II)–cyanide and zinc(II)–ammonia systems, the change from $\bar{n} = 1$ to 3 occurs over a ligand concentration range of ~4.1-fold and ~3.5-fold, respectively. Each is smaller than the 9-fold change for the statistical model.

A useful model for these systems is that proposed by Monod, Wyman, and Changeux (24) for the cooperative binding of oxygen by hemoglobin. In this model, the protein (e.g., hemoglobin) exists in two different forms that can bind ligand (oxygen); cooperativity is manifested if the form of the protein that is less stable in the absence of ligand binds the ligand more strongly than does the more stable form. The difference of structures of aquazinc(II) ion and the tetraligated complexes encourages application of this model.

For a system containing a dominant octahedral aquazinc(II) ion that also can form tetrahedral species, the equations relating the empirical overall equilibrium constants β_n with this model are (20)

$$\beta_1 = q_o(6 + 4k_0 r) \qquad \beta_2 = q_o^2(3 + 12\alpha_o + 6k_0 r^2 \alpha_t)$$

$$\beta_3 = q_o^3(12\alpha_o^2 + 8\alpha_o^3 + 4k_0 r^3 \alpha_t^3)$$

$$\beta_4 = q_o^4(3\alpha_o^4 + 12\alpha_o^5 + k_0 r^4 \alpha_t^6) \qquad \beta_5 = 6q_o^5 \alpha_o^8$$

$$\beta_6 = q_o^6 \alpha_o^{12}$$

The numerical coefficients in these equations are statistical factors, and the parameters used in this formulation are as follows: q_o, the statistically corrected equilibrium constant for formation of octahedral $Zn(OH_2)_5X$ from $Zn(OH_2)_6^{2+}$; k_0, the relative concentration of the tetrahedral aquametal ion ($k_0 = [Zn(OH_2)_4^{2+}]/[Zn(OH_2)_6^{2+}]$); r, the ratio of the statistically corrected equilibrium constants for formation of $Zn(OH_2)_3X$ and $Zn(OH_2)_5X$ (i.e., $r = q_t/q_o$ if q_t is the statistically corrected equilibrium constant for formation of $Zn(OH_2)_3X$ from $Zn(OH_2)_4^{2+}$); α_t and α_o, the factors in an equilibrium constant arising because the reaction involves the placement of ligands at adjacent sites of the coordination polyhedron (α_t for tetrahedral coordination and α_o for octahedral coordination). These equations reveal the basis for cooperativity in this model. With $k_0 \ll 1$ and $r \gg 1$, the contribution of the form that is unimportant for the hydrated ion and the monoligand complex can become important, even highly dominant, for the higher complexes, because of the increasing power to which r is raised in equations for β_n ($n = 3, 4$). (If k_0 is not small compared with 1, a factor $(1 + k_0)^{-1}$ should be included on the right-hand side of the equation for each β_n.)

Table II. Parameters That Correlate $\bar{n}([X])$ for Zn^{2+}/NH_3 and Zn^{2+}/CN^- Complex Ion Systems

	Zn^{2+}/NH_3	Zn^{2+}/CN^-
k_0	0.003	0.003
r_o	40	150
q_0	44.1	2.54×10^4
α_t^1	0.69	0.42
$\sigma_{rms}(\bar{n})$	0.018	0.052

Note: The value of α_o plays little role in correlating the data, since species with two or more ligands are predominantly tetrahedral. For these calculations it was assumed that $\alpha_o = \alpha_t$.

Table II summarizes the parameters that correlate the data for the zinc(II) complexes under consideration. For the model to be considered successful, a particular value of k_0 must correlate the data for all systems involving a particular metal ion. For the zinc(II) systems, the value 0.003 correlates data for both ammine and cyanide complexes. With the parameters presented in Table II, one can calculate the relative concentrations of tetrahedral and octahedral species with each particular number of ligands. The results of these calculations are presented in Table III.

Table III. Relative Concentration of Species With Tetrahedral and Octahedral Coordination

$$k_n = \frac{[Zn(OH_2)_{4-n}X_n]}{[Zn(OH_2)_{6-n}X_n]}$$

	X = NH$_3$	X = CN$^-$
k_0	0.003	0.003
k_1	0.080	0.30
k_2	1.8	21.
k_3	30	1.1×10^3
k_4	3.2×10^2	1.1×10^6

For each of these series of complexes, octahedral species are more important than tetrahedral species for $n = 0$ and 1, but the opposite is true for $n = 2, 3$, and 4. The model does not necessarily dictate an abrupt transition from octahedral to tetrahedral coordination at a particular value of n, but we see in these derived results that the transition in forming $Zn(OH_2)_2X_2$ from $Zn(OH_2)_5X$ is relatively abrupt. For the ammine system, the fraction of tetrahedral species increases in this step from 0.08 to 0.64, and for the cyanide system the increase is from 0.23 to 0.95. This model allows for a transition to octahedral species, $Zn(OH_2)X_5$ and ZnX_6, but this would occur at concentrations of ligand outside of the range studied.

One would expect other properties of these systems to manifest the reduction in coordination number. In particular, a reaction converting $Zn(OH_2)_5X$ to $Zn(OH_2)_2X_2$ should have a less negative value of ΔH and a more positive value of ΔS^0 than expected if no change of coordination number occurred. Highly accurate calorimetric data would be needed to shed light on this question, however.

For the zinc(II) systems just considered, the two forms of aqua metal ion have different coordination geometries. For other cooperative complex ion systems, the two forms are the high-spin and low-spin forms with the same or with different coordination geometries, e.g., the cyanide complexes of iron(II), iron(III), and nickel(II) (20).

Kinetics and mechanisms

The Reaction of Uranium(IV) with Plutonium(VI)

T. W. Newton (25) in his study of this reaction raises an interesting point about ambiguity in relating the empirical rate constant to the rate constant for the slow reaction step. The ambiguity arises in this system because the identity of the reaction step that follows the slow step influences our interpretation of the empirical rate constant.

The net chemical change is

$$U^{4+} + 2PuO_2^{2+} + 2H_2O = UO_2^{2+} + 2PuO_2^+ + 4H^+,$$

and the species shown in this equation are dominant at the high-acidity extreme of the study ($[H^+] = 1.50$ mol dm^{-3}). At the low-acidity extreme ($[H^+] = 0.10$ mol dm^{-3}), uranium(IV) is converted partially to UOH^{3+}, and this will be taken into account in interpretation of the acidity dependence of the rate constant. Runs at a particular acidity are governed by the second-order rate law

$$-\frac{1}{2}\frac{d[PuO_2^{2+}]}{dt} = k'[PuO_2^{2+}][U^{IV}]$$

where $[U^{IV}] = [U^{4+}] + [UOH^{3+}]$. Two reasonable mechanisms involving UO$_2^+$ as an intermediate are suggested by this rate law. The initial step for each is the same (since we are considering at this point only runs at a particular acidity, these reaction steps are not balanced with respect to hydrogen ion, and the species of uranium(IV) is not specified):

$$PuO_2^{2+} + U^{IV} \xrightarrow{k_1} PuO_2^+ + UO_2^+ \tag{1}$$

To complete the reaction stoichiometry, this reaction can be followed by either

$$PuO_2^{2+} + UO_2^+ \xrightarrow{k_2} PuO_2^+ + UO_2^{2+} \qquad (2)$$

or

$$2\,UO_2^+ \xrightarrow{k_3} U^{IV} + UO_2^{2+} \qquad (3)$$

Thus the overall reaction is the sum (1) + (2), in which case the empirical rate constant k' is equal to k_1; or the overall reaction is the sum $[2 \times (1)] + (3)$, in which case k' is equal to $1/2\,k_1$. With no data to support one or the other of these mechanisms, interpretation of the empirical rate constant in terms of the rate of reaction (1) is uncertain. In particular, if the temperature coefficient of the reaction rate were known, calculation of the standard entropy change for activation in reaction (1) would be ambiguous. The values of $^{\ddagger}\Delta S^0$ calculated under the two assumed relationships $k_1 = k'$ and $k_1 = 2k'$ would differ by R ln 2 (or 5.8 JK^{-1} mol^{-1}).

Although kinetic behavior arising from the simultaneous occurrence of reactions (2) and (3) at similar rates was not observed, the consequences of such circumstances can be explored. The steady-state method gives the quadratic equation

$$2k_3[UO_2^+]^2 + k_2[PuO_2^{2+}][UO_2^+] - k_1[PuO_2^{2+}][U^{IV}] \cong 0$$

that has the positive root

$$[UO_2^+] = \frac{k_2[PuO_2^{2+}]}{4k_3}\left\{\left(1 + \frac{8k_1k_3[U^{IV}]}{k_2^2[PuO_2^{2+}]}\right)^{1/2} - 1\right\}$$

Substitution of this concentration into the rate law

$$-\frac{d[U^{IV}]}{dt} = -\frac{1}{2}\frac{d[PuO_2^{2+}]}{dt} = \frac{1}{2}k_1[PuO_2^{2+}][U^{IV}] + \frac{1}{2}k_2[PuO_2^{2+}][UO_2^+]$$

gives

$$-\frac{d[U^{IV}]}{dt} = \frac{1}{2}k_1[PuO_2^{2+}][U^{IV}] + \frac{k_2^2[PuO_2^{2+}]^2}{8k_3}\left\{\left(1 + \frac{8k_1k_3[U^{IV}]}{k_2^2[PuO_2^{2+}]}\right)^{1/2} - 1\right\}$$

The limiting forms of this rate law are those already presented:

$$-\frac{d[U^{IV}]}{dt} = k_1[PuO_2^{2+}][U^{IV}]$$

and

$$-\frac{d[U^{IV}]}{dt} = \frac{1}{2}k_1[PuO_2^{2+}][U^{IV}]$$

Clearly, however, the apparent reaction orders for uranium(IV) and plutonium(VI) would deviate from unity under conditions that do not correspond to one or the other limiting case.

It is easy enough to see the basis for the twofold difference in the second-order rate constant, depending upon which reaction step follows the initial step. If reaction (2) follows reaction (1), the rate of disappearance of uranium(IV) in step 1 is its rate of reaction. If reaction (3) follows reaction (1), the rate of disappearance of uranium(IV) in step (1) is a factor of two times its rate of reaction, because reaction (3) regenerates one uranium(IV) for every two that are consumed in reaction (1).

It is at this point that we may speculate regarding which possible reaction does, in fact, follow the slow step. Reaction of the predominant forms of uranium(V) and plutonium(VI) is simply an electron transfer:

$$PuO_2^{2+} + UO_2^+ \xrightarrow{k_2} PuO_2^+ + UO_2^{2+}$$

but the disproportionation of uranium(V)

$$2UO_2^+ + 4H^+ \rightarrow UO_2^{2+} + U^{4+} + 2H_2O$$

involves proton shuffling and changes of metal ion—oxide ion bonding. Thus it is reasonable to assume that it is reaction (2) that follows the slow step. Some support for this is provided by the dependence upon acidity of the empirical second-order rate constant.

The empirical rate constant for this reaction increases with a decrease in the concentration of hydrogen ion, and the dependence reveals additional aspects of the reaction mechanism. The hydrolysis constant for uranium(IV) ($K_a(U^{4+}) = 0.024$ mol dm^{-3}) indicates that hydrolysis occurs to an appreciable extent at the lowest acidities studied. With K_a known, this should be taken into account prior to interpretation of the hydrogen ion dependence of the rate. We do this by equating two expressions of the right-hand side of the rate law

$$k'[PuO_2^+][U^{IV}] = k''[PuO_2^{2+}][U^{4+}]$$

It is k'', the rate constant for a rate law that involves the concentrations of species, for which we wish to interpret the dependence upon acidity. Thus

$$k'' = k' \times \frac{[U(IV)]}{[U^{4+}]} = k' \times \frac{K_a(U^{4+}) + [H^+]}{[H^+]}$$

Figure 10. The dependence of k'' for the reaction of uranium(IV) and plutonium(VI) upon the hydrogen ion concentration

$$k'' = k' \times \frac{K_a(U^{4+}) + [H^+]}{[H^+]}$$

where k' is the empirical second-order rate constant.

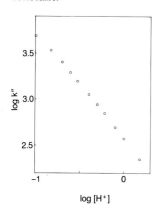

Figure 10 is a plot of log k'' versus log $[H^+]$. The slope of this line is the apparent order with respect to hydrogen ion, and it decreases mildly with increasing acidity, being -1.0 at the lowest acidity ($[H^+] = 0.10$ mol dm^{-3}) and -1.3 at the highest acidity ($[H^+] = 1.5$ mol dm^{-3}). A decreasing reaction order for hydrogen ion with increasing concentration of hydrogen ion indicates a sum of terms in the denominator of the rate law, with the concentration of hydrogen ion raised to different powers in each term. The form of the rate law suggested by Figure 10 is

$$-\frac{1}{2}\frac{d[PuO_2^{2+}]}{dt} = k\frac{[PuO_2^{2+}][U^{4+}]}{[H^+] + \kappa[H^+]^2}$$

Thus the empirical rate constant corrected for hydrolysis of uranium(IV), k'', is

$$k'' = k/([H^+] + \kappa[H^+]^2)$$

Rearrangement yields

$$\frac{1}{k''[H^+]} = \frac{1}{k} + \frac{\kappa[H^+]}{k}$$

which leads to the expectation that a plot of $(k''[H^+])^{-1}$ versus $[H^+]$ will be linear. This expectation is confirmed, as is shown in Figure 11.

Figure 11. Evaluation of k and κ for reaction of U^{4+} with PuO_2^{2+} (see text).

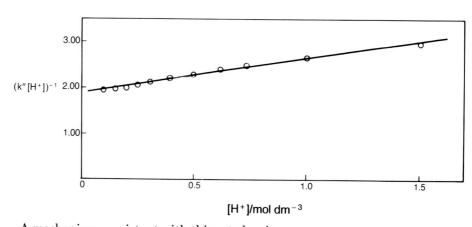

A mechanism consistent with this rate law is

$$U^{4+} + H_2O \overset{K_a}{\rightleftharpoons} UOH^{3+} + H^+$$

$$UOH^{3+} + PuO_2^{2+} \underset{k'_{-1}}{\overset{k'_1}{\rightleftharpoons}} OUOPuO^{4+} + H^+$$

$$H_2O + OUOPuO^{4+} \overset{k'_2}{\rightarrow} UO_2^+ + PuO_2^+ + 2H^+$$

followed by the rapid reaction of PuO_2^{2+} and UO_2^+. If this is the mechanism, the parameter k is identified as $(K_a(U^{4+}) \times k'_1)$, and κ is identified as k'_{-1}/k'_2. It is reasonable to propose that the intermediate OUOPuO^{4+}, the product of the first step, persists long enough to have alternate fates, the k'_{-1} step that involves attack by acid and the k'_2 step that involves coordination shell alterations. With kinetically inert metal ions [e.g., Cr^{3+}], dimeric metal species (26) (($H_2O)_5CrONpO^{4+}$, a species containing Cr(III) and Np(V), formed from Cr^{2+} and NpO_2^{2+}) can be isolated.

A reaction displaying some of the characteristics just discussed is that of iodine and tetrathionate ion (27):

$$S_4O_6^{2-} + 7\,I_2 + 10\,H_2O = 4\,SO_4^{2-} + 14\,I^- + 20\,H^+$$

The rate of this complex reaction under two sets of concentration conditions (one with relatively high concentrations of iodide ion and tetrathionate ion and the other with a low concentration of tetrathionate ion and no iodide ion present initially) is governed by the same rate law

$$-\frac{d[S_4O_6^{2-}]}{dt} = \frac{k[S_4O_6^{2-}][I_2]}{[I^-]}$$

Under the conditions of low concentrations of tetrathionate ion and iodide ion, the empirical rate constant is approximately two times larger than it is under the other set of concentration conditions. Awtrey and Connick (27) attribute this to the phenomenon just discussed; there are alternate reaction pathways after the rate-determining step.

A Complex Acidity Dependence

The reaction of chloramine with glycine exhibits a complex dependence upon the concentration of hydrogen ion that is resolved nicely when account is taken of the protonation equilibria of the reactant species. The kinetic study of this reaction by Snyder and Margerum (28) extended from pH 1.7 to 10.4, in which range there is appreciable stability for the two forms of chloramine ($ClNH_2$ and $ClNH_3^+$) and the three forms of glycine ($H_3\overset{+}{N}CH_2CO_2H$, $H_3\overset{+}{N}CH_2CO_2^-$, and $H_2NCH_2CO_2^-$). At each particular acidity, the rate law is

$$-\frac{dC_A}{dt} = k'C_A C_G$$

where $C_A = [ClNH_2] + [ClNH_3^+]$, and $C_G = [H_3\overset{+}{N}CH_2CO_2H] + [H_3\overset{+}{N}CH_2CO_2^-] + [H_2NCH_2CO_2^-]$. (The non-zwitterion form of the neutral species is a very minor species.) The dependence of k' upon the acidity is shown in Figure 12.

Figure 12. The dependence of k' for the reaction of chloramine with glycine upon the hydrogen ion concentration. (k' is empirical second-order rate constant.)

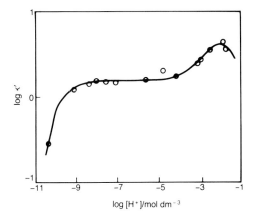

As in our discussion of the hydrogen ion dependence of the plutonium(VI)–uranium(IV) reaction, the first step in unraveling this dependence is expression of the rate law in terms of concentration of species. For the neutral species

$$-\frac{dC_A}{dt} = k[ClNH_2][H_3\overset{+}{N}CH_2CO_2^-]$$

the rate constant k is related to k' by the equation

$$k = k' \times \frac{C_A}{[\text{ClNH}_2]} \times \frac{C_G}{[\text{H}_3\overset{+}{\text{N}}\text{CH}_2\text{CO}_2^-]}$$

$$= k' \times \left(\frac{[\text{H}^+] + K(\text{AH}^+)}{K(\text{AH}^+)}\right) \times \left(\frac{[\text{H}^+]^2 + K(\text{GH}_2^+)[\text{H}^+] + K(\text{GH}_2^+)K(\text{GH})}{K(\text{GH}_2^+)[\text{H}^+]}\right)$$

in which $K(\text{AH}^+)$ is the acid dissociation constant for chlorammonium ion, and $K(\text{GH}_2^+)$ and $K(\text{GH})$ are the successive acid dissociation constants for glycine cation. The data points of Figure 12 are transformed to those of Figure 13, a plot of log k versus pH, by the equation given above. We see in Figure 13 that the reaction order for hydrogen ion increases with an increase in the concentration of hydrogen ion. This trend, the opposite of that observed in the uranium(IV)–plutonium(VI) reaction, indicates a sum of terms involving different powers of the concentration of hydrogen ion. The complete rate law

Figure 13. The dependence of k for the reaction of chloramine with glycine upon the hydrogen ion concentration. (k is the second-order rate constant for reaction of ClNH_2 with $\text{H}_3\overset{+}{\text{N}}\text{CH}_2\text{CO}_2^-$.)

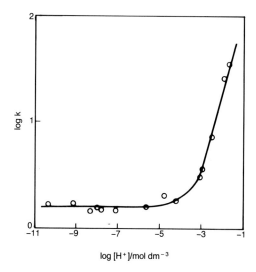

$$-\frac{dC_A}{dt} = (1.6 \text{ dm}^3 \text{ mol}^{-1} \text{ s}^{-1} + 1.8 \times 10^3 \text{ dm}^6 \text{ mol}^{-2} \text{ s}^{-1} [\text{H}^+])$$

$$\times [\text{ClNH}_2][\text{H}_3\overset{+}{\text{N}}\text{CH}_2\text{CO}_2^-]$$

indicates activated complexes having net charges 0 and +1. Now Figure 12 is easily understood. At the lowest concentration of hydrogen ion studied, chloramine is in the neutral form and glycine is anionic. One proton must be added to produce the neutral activated complex; the empirical rate constant is proportional to the concentration of hydrogen ion raised to the first power. In the range of acidity from pH 8 to pH 4, the predominant form of each reactant is neutral, and the neutral activated complex is formed with neither addition nor subtraction of a proton; the empirical rate constant is independent of the concentration of hydrogen ion. The cationic activated complex becomes important at $[\text{H}^+] > 10^{-4}$ mol dm^{-3}, and since the neutral forms of the reactants are still the predominant forms, a proton must be added to form the activated complex; the empirical rate constant again increases with increasing acidity. But not for long. At pH ~2.4 glycine is half converted to a cationic species, and at pH ~1.5 chloramine is half converted to a cationic species. Thus over a narrow range of acidity the reaction order in hydrogen ion goes from 0 to +1 to 0 to −1. More studies would have to be made at high acidity to disclose whether or not there is an important activated complex having charge 2+.

The Reaction of Thallium(III) and Iron(II): Integration of a Complex Rate Law
The reaction of thallium(III) and iron(II) in acidic aqueous solution

$$Tl^{3+} + 2\,Fe^{2+} = Tl^+ + 2\,Fe^{3+}$$

is governed by the rate law (*29, 30*)

$$-\frac{d[Tl^{3+}]}{dt} = \frac{k[Tl^{3+}][Fe^{2+}]^2}{[Fe^{2+}] + \kappa[Fe^{3+}]}$$

This rate law is consistent with the mechanism involving Tl^{2+} as an unstable intermediate:

$$Tl^{3+} + Fe^{2+} \underset{k_{-1}}{\overset{k_1}{\rightleftharpoons}} Tl^{2+} + Fe^{3+}$$

$$Tl^{2+} + Fe^{2+} \overset{k_2}{\rightarrow} Tl^+ + Fe^{3+}$$

The empirical parameters are identified with the rate constants of this mechanism: $k = k_1$, and $\kappa = k_{-1}/k_2$. Now we will discuss a number of points illustrated by this system.

The usual steady-state derivation for this mechanism involves equating $d[Tl^{2+}]/dt$ to 0 to yield an equation

$$[Tl^{2+}] = \frac{k_1[Fe^{2+}][Tl^{3+}]}{k_2[Fe^{2+}] + k_{-1}[Fe^{3+}]}$$

for the steady-state concentration of thallium(II). If one does not equate $d[Tl^{2+}]/dt$ to 0, the equation for $[Tl^{2+}]$ is

$$[Tl^{2+}] = \frac{k_1[Fe^{2+}][Tl^{3+}]}{k_2[Fe^{2+}] + k_{-1}[Fe^{3+}]} - \frac{1}{k_2[Fe^{2+}] + k_{-1}[Fe^{3+}]}\frac{d[Tl^{2+}]}{dt}$$

Substitution of this relationship into the equation for $-d[Tl^{3+}]/dt$ yields

$$-\frac{d[Tl^{3+}]}{dt} = \frac{k_1 k_2 [Tl^{3+}][Fe^{2+}]^2}{k_2[Fe^{2+}] + k_{-1}[Fe^{3+}]} + \frac{k_{-1}[Fe^{3+}]}{k_2[Fe^{2+}] + k_{-1}[Fe^{3+}]}\frac{d[Tl^{2+}]}{dt}$$

This formulation shows that the steady-state rate law depends for its validity not on $d[Tl^{2+}]/dt$ being 0, but on the less stringent requirement $|d[Tl^{2+}]/dt| \ll |d[Tl^{3+}]/dt|$.

This empirical rate law can be used to illustrate the integration of a complex rate law by the method of partial fractions (*31*). If thallium(III) is the limiting reagent, and if allowance is made for having iron(III) present at zero time, the rate law can be written

$$-\frac{d[Tl^{3+}]}{dt} = \frac{k[Tl^{3+}]([Fe^{2+}]_\infty + 2[Tl^{3+}])^2}{([Fe^{2+}]_\infty + 2[Tl^{3+}]) + \kappa([Fe^{3+}]_\infty - 2[Tl^{3+}])}$$

since $[Fe^{2+}] = [Fe^{2+}]_\infty + 2[Tl^{3+}]$, and $[Fe^{3+}] = [Fe^{3+}]_\infty - 2[Tl^{3+}]$. Rearrangement to

$$\frac{[Fe^{2+}]_\infty + \kappa[Fe^{3+}]_\infty + 2[Tl^{3+}](1-\kappa)}{[Tl^{3+}]([Fe^{2+}]_\infty + 2[Tl^{3+}])^2} \times d[Tl^{3+}] = -k\,dt$$

is the starting point for application of the method of partial fractions. The coefficient of $d[Tl^{3+}]$ can be expressed as

$$\frac{A}{[Tl^{3+}]} + \frac{B}{[Fe^{2+}]_\infty + 2[Tl^{3+}]} + \frac{C}{([Fe^{2+}]_\infty + 2[Tl^{3+}])^2}$$

or

$$\frac{A([Fe^{2+}]_\infty + 2[Tl^{3+}])^2 + B[Tl^{3+}]([Fe^{2+}]_\infty + 2[Tl^{3+}]) + C[Tl^{3+}]}{[Tl^{3+}]([Fe^{2+}]_\infty + 2[Tl^{3+}])^2}$$

If $A([Fe^{2+}]_\infty + 2[Tl^{3+}])^2 + B[Tl^{3+}]([Fe^{2+}]_\infty + 2[Tl^{3+}]) + C[Tl^{3+}]$ is to be equal to $[Fe^{2+}]_\infty + \kappa[Fe^{3+}]_\infty + 2[Tl^{3+}](1-\kappa)$ for all concentrations of thallium(III), coefficients for each particular power of the concentration of thallium(III) must be equal in these two expressions. This allows evaluation of A, B, and C:

$$A = -\frac{B}{2} = \frac{[Fe^{2+}]_\infty + \kappa[Fe^{3+}]_\infty}{[Fe^{2+}]_\infty^{\,2}}$$

$$C = -2\kappa\left(1 + \frac{[Fe^{3+}]_\infty}{[Fe^{2+}]_\infty}\right)$$

Substitution of these relationships and $d[Tl^{3+}] = \frac{1}{2}d[Fe^{2+}]$ and $[Fe^{2+}] = [Fe^{2+}]_\infty + 2[Tl^{3+}]$ gives for the integrated rate law

$$\frac{[Fe^{2+}]_\infty + \kappa[Fe^{3+}]_\infty}{[Fe^{2+}]_\infty^2} \ln\frac{[Tl^{3+}][Fe^{2+}]_0}{[Tl^{3+}]_0[Fe^{2+}]}$$
$$+ \kappa\left(1 + \frac{[Fe^{3+}]_\infty}{[Fe^{2+}]_\infty}\right)\left(\frac{1}{[Fe^{2+}]} - \frac{1}{[Fe^{2+}]_0}\right) = -kt$$

This equation can be rearranged to be amenable to graphical evaluation of k and κ

$$\frac{\left(1 + \frac{[Fe^{3+}]_\infty}{[Fe^{2+}]_\infty}\right)\left(\frac{1}{[Fe^{2+}]} - \frac{1}{[Fe^{2+}]_0}\right)}{\ln\frac{[Tl^{3+}][Fe^{2+}]_0}{[Tl^{3+}]_0[Fe^{2+}]}} = \frac{-k}{\kappa}\frac{t}{\ln\frac{[Tl^{3+}][Fe^{2+}]_0}{[Tl^{3+}]_0[Fe^{2+}]}} - \frac{[Fe^{2+}]_\infty + \kappa[Fe^{3+}]}{\kappa[Fe^{2+}]_\infty^2}$$

A plot of the function on the left-hand side of the equation versus

$$t\Big/\left(\ln\frac{[Tl^{3+}][Fe^{2+}]_0}{[Tl^{3+}]_0[Fe^{2+}]}\right) \text{ yields } -k/\kappa \text{ as the slope and}$$

$-([Fe^{2+}]_\infty + \kappa[Fe^{3+}]_\infty)/\kappa[Fe^{2+}]_\infty^2$ as the intercept.

The integrated rate law has terms in $\ln [Tl^{3+}]$, $\ln [Fe^{2+}]$, and $[Fe^{2+}]^{-1}$, each of which by itself would arise from a reaction that is first order in $[Tl^{3+}]$, first order in $[Fe^{2+}]$, and second order in $[Fe^{2+}]$, respectively. It is the concentration dependences in the numerator of the rate law that determine the types of terms in the integrated rate law. It is the multiterm denominator of the rate law that determines the coefficients of terms in the integrated rate law. Thus other rate laws with a numerator $k[Tl^{3+}][Fe^{2+}]^2$ would give an integrated rate law having the same form as that already derived, but the coefficients of the terms would be different. Thus a rate law with the same numerator but a denominator $[Fe^{2+}] + \kappa[Tl^+]$, which arises if Fe^{4+} is the unstable intermediate produced in the first step, would not in a single run be distinguishable from the rate law we have just considered. But runs with varying initial concentrations of thallium(I) and iron(III) would distinguish the rate laws.

Many postulated reaction intermediates defy attempts to prepare them at detectable levels. The species Tl^{2+} is not such a species. It has been prepared by the oxidation of thallium(I) by hydroxyl radicals and the reduction of thallium(III) by atomic hydrogen (30). Its reaction with iron(II), the second step of the mechanism just considered, has been studied directly. This allows evaluation of k_2 which, with the values of $\kappa(k_{-1}/k_2)$ and with k_1 evaluated from analysis of the kinetic data on the reaction of thallium(III) with iron(II), gives for 25 °C (30) and $[H^+] = 1.1$ mol dm^{-3}: $k_1 = (1.39 \pm 0.02) \times 10^{-2}$ dm^3 mol^{-1} s^{-1}, $k_{-1} = 3.4 \times 10^5$ dm^3 mol^{-1} s^{-1}, and $k_2 = 6.7 \times 10^6$ dm^3 mol^{-1} s^{-1}.

The Decomposition of Ozone Catalyzed by Dinitrogenpentaoxide

The previous discussion does not prepare us for the form of the steady-state rate law generated by a mechanism in which each of two intermediates produced in a rapid equilibrium goes on to products by separate reactions. Such a situation arises in the decomposition of ozone catalyzed by dinitrogenpentaoxide (32). The net chemical change is

$$2\, O_3 = 3\, O_2$$

and each of the reaction orders for ozone and dinitrogenpentaoxide is two-thirds:

$$-\frac{1}{2}\frac{d[O_3]}{dt} = k[O_3]^{2/3}[N_2O_5]^{2/3}$$

From this rate law one cannot discern a rational transition-state composition. A mechanism accounting for these reaction orders is

$$N_2O_5 \underset{}{\overset{K_1}{\rightleftharpoons}} NO_2 + NO_3 \quad \text{rapid equilibrium} \tag{1}$$

$$NO_2 + O_3 \xrightarrow{k_2} NO_3 + O_2 \tag{2}$$

$$2\,NO_3 \xrightarrow{k_3} 2\,NO_2 + O_2 \tag{3}$$

The net chemical change is two times reaction (2) plus reaction (3). Combining equations for the equilibrium condition in reaction (1) and the steady-state conditions for NO_2 and NO_3 in reactions (2) and (3) gives

$$[NO_2] = \left(\frac{2k_3 K_1^2 [N_2O_5]^2}{k_2 [O_3]}\right)^{1/3}$$

which yields, for the rate of reaction (2), the ozone-consuming reaction:

$$-\frac{d[O_3]}{dt} = K_1^{2/3} k_2^{2/3} (2\,k_3)^{1/3} [N_2O_5]^{2/3} [O_3]^{2/3}$$

In this mechanism a rapid equilibrium produces two different unstable intermediates, each of which then reacts separately in unidirectional slow steps. This type of mechanism gives a rate law with concentration dependences that do not simply disclose the composition of activated complexes. This mechanism does not have a rate-determining step.

It should be noted that the rate of this reaction is proportional to the geometric mean of the rates of reactions (2) and (3), with reaction (2) having a twofold greater weight than reaction (3).

The Rate Laws for Reversible Reactions (33)

The number of reactions amenable to kinetic study that proceed to a measurable equilibrium at ordinary concentration conditions is limited, but the relationship between the concentration dependences of the forward and reverse rate laws is an appropriate topic for discussion.

The Disproportionation Reactions of Nitrous Acid and Nitrogen Dioxide

The disproportionation reactions of nitrous acid and nitrogen dioxide are ideal examples with which to illustrate some aspects of this subject. The potential diagram (Latimer diagram) for nitrogen (oxidation states 2+ through 5+) for acidic solution (34):

$$NO_3^- \xrightarrow{0.80V} NO_2(g) \xrightarrow{1.06V} HONO \xrightarrow{1.04V} NO(g)$$

with 1.05V (NO_2 to NO) and 0.93V (NO_3^- to HONO) couples indicated.

reveals that each of the disproportionation reactions

$$3\,NO_2(g) + H_2O = 2\,H^+ + 2\,NO_3^- + NO(g) \tag{1}$$

$$3\,HONO = H^+ + NO_3^- + 2\,NO(g) + H_2O \tag{2}$$

proceeds to a measurable equilibrium under realizable concentration conditions. Reaction (1), a step in the production of nitric acid, is the predominant reaction when nitrogen dioxide is added to nitric acid; reaction (2) is the predominant decomposition reaction for nitrous acid in dilute acidic solution. Kinetic studies of each of these reactions are consistent with a mechanism consisting of the same reaction steps, a rapid equilibrium reaction (3) and a slow step reaction (4):

$$2\,HONO \underset{}{\overset{K_3}{\rightleftharpoons}} NO_2(g) + NO(g) + H_2O \tag{3}$$

$$2\,NO_2(g) + H_2O \underset{k_{-4}}{\overset{k_4}{\rightleftharpoons}} H^+ + NO_3^- + HONO \tag{4}$$

Reaction (1) is reaction (3) plus *two times* reaction (4), and reaction (2) is *two times* reaction (3) plus reaction (4). In reaction (1) the unstable intermediate is nitrous acid, and in reaction (2) the unstable intermediate is nitrogen dioxide. The empirical rate laws (with activity coefficient factors omitted [35, 36]) that suggest this mechanism are

$$\frac{1}{2}\frac{d[NO_3^-]}{dt} = k_{1f} P_{NO_2}^2 - k_{1r}[H^+][NO_3^-] P_{NO}^{1/2} P_{NO_2}^{1/2} \tag{5}$$

$$\frac{d[NO_3^-]}{dt} = k_{2f}\frac{[HONO]^4}{P_{NO}^2} - k_{2r}[H^+][NO_3^-][HONO] \tag{6}$$

with $k_{1f} = \frac{1}{2}k_4$, $k_{1r} = \frac{1}{2}k_{-4} K_3^{-1/2}$, $k_{2f} = K_3^2 k_4$, and $k_{2r} = k_{-4}$. The important feature of these rate laws to be noted at this point is that equating to zero the rate given by Equation (5) yields the square root of the equilibrium constant for reaction (1), and equating to zero the rate given by Equation (6) yields the equilibrium constant for reaction (2):

$$\frac{1}{2} k_4 P_{NO_2}^2 - \frac{1}{2} k_{-4} K_3^{-1/2} [H^+][NO_3^-] P_{NO_2}^{1/2} P_{NO_2}^{1/2} = 0$$

yields $\quad \dfrac{[H^+][NO_3^-] P_{NO}^{1/2}}{P_{NO_2}^{3/2}} = \dfrac{k_4}{k_{-4}} K_3^{1/2} = K_1^{1/2}$

and

$$K_3^2 k_4 \frac{[HONO]^4}{P_{NO}^2} - k_{-4}[H^+][NO_3^-][HONO] = 0$$

yields $\quad \dfrac{[H^+][NO_3^-] P_{NO}^2}{[HONO]^3} = \dfrac{k_4}{k_{-4}} K_3^2 = K_2$

Thus reaction (1) provides an example of a rate law for a reversible reaction that upon being equated to zero yields not the conventional equilibrium constant equation, that involving smallest integral exponents, but rather the square root of the conventional equilibrium constant equation. On the other hand, the rate law for reaction (2), equation (6), yields the conventional equilibrium constant. The key to the difference between these two situations is the factor by which the equation for the slow step, reaction (4), must be multiplied in the summation of steps that is the net change. In the summation to obtain reaction (1), equation (4) is multiplied by *two*; in the summation to obtain reaction (2), equation (4) is multiplied by *one*. That is, the stoichiometric number[1] for the slow step is *two* for reaction (1) and is *one* for reaction (2).

As shown in this example, the rate law for a reversible reaction involves rate constants for the forward and reverse of the slow step, *each raised to the first power*. The rate law may involve ratios of additional rate constants, and these ratios may be raised to powers other than 1. In these examples, we see K_3 (a ratio of rate constants) raised to the one-half power and to the second power.

The equilibrium constant expression derived by equating the rate to zero involves the first power of the quotient of rate constants for the forward and reverse of the slow step. And a corollary of this is that the balanced chemical equation that goes with the rate law for a reversible reaction is that which is obtained as a sum of reaction steps including *one* times the rate-determining step. If to obtain the conventional chemical equation for the reaction, that with smallest integral coefficients, the equation for the slow step must be multiplied by a factor other than 1, the equilibrium constant equation based upon the rate law will be the conventional one raised to a power other than unity; it is 1/2 in the present example. This point has been made by Horiuti and Nakamura (37), who give as the relationship between the equilibrium constant and empirical rate constants for forward and reverse reactions:

---

[1] The term stoichiometric number is used here as in Horiuti and Nakamura (37). The stoichiometric number for a reaction step depends, therefore, upon coefficients in the overall reaction, as will be illustrated.

$$\frac{k_f}{k_r} = K^{1/\nu} \qquad (7)$$

in which ν is the stoichiometric number for the rate-determining step. Thus for reaction (1), $k_f/k_r = K^{1/2}$, and for reaction (2), $k_f/k_r = K$. One can make the stoichiometric number for the slow step equal to 1 simply by adjusting the coefficients in the equation for the overall reaction. If reaction (1) is multiplied by 1/2, the stoichiometric number for the slow step becomes 1.

This discussion is, of course, background for the question of predicting concentration dependences of a reverse rate law given the concentration dependences of a forward rate law. Clearly the mechanism must be known, in particular the stoichiometric number of the slow step, to make a certain prediction. In the absence of definitive data, the cautious prediction probably is that based upon the assumption that the stoichiometric number of the slow step is 1. Other than the example before us, there are few systems for which the cautious prediction would fail.

The Reaction of Tris(triphenylphosphine)platinum(0) with 1-phenyl-1-propyne

Another facet of this subject can be illustrated by the ligand exchange of tris(triphenylphosphine)platinum(0) studied by Halpern and Weil (38). The rate of the reversible reaction

$$Pt(PPh_3)_3 + CH_3CCPh = Pt(PPh_3)_2(CH_3CCPh) + PPh_3$$

studied in benzene at 25 °C, is governed by the rate law:

$$-\frac{d[A]}{dt} = [A][B]\left(0.050 \text{ dm}^3 \text{ mol}^{-1} \text{ s}^{-1} + \frac{0.91 \text{ s}^{-1}}{250[D] + [B]}\right)(1 - 1.43\,Q)$$

in which $A = Pt(PPh_3)_3$, $B = CH_3CCPh$, $C = Pt(PPh_3)_2(CH_3CCPh)$, $D = PPh_3$, and $Q = [C][D]/([A][B])$. This rate law is consistent with concurrent pathways, a direct associative pathway (a reaction step with the stoichiometry of the overall reaction) with a second-order rate constant for the forward reaction of 0.050 dm^3 mol^{-1} s^{-1} and a dissociative pathway:

$$Pt(PPh_3)_3 \rightleftarrows Pt(PPh_3)_2 + PPh_3$$

$$Pt(PPh_3)_2 + CH_3CCPh \rightleftarrows Pt(PPh_3)_2(CH_3CCPh)$$

The intermediate $Pt(PPh_3)_2$ has a formation rate constant from $Pt(PPh_3)_3$ of 0.91 s^{-1} and, in being competed for by the two ligands, discriminates in favor of triphenylphosphine by a factor of 250. The negative term in the rate law corresponds to the reverse reaction, and the numerical factor in this term is the reciprocal of the equilibrium constant, K^{-1} ($K = [C][D]/[A][B] = 0.70$). Halpern and Weil make the interesting observation that a pseudo-first-order study (($[A] + [C]) \ll [B], [D]$) under conditions where equilibrium lies far to the right ($[B] = 0.30$ mol dm^{-3}, $[D] = 1.0 \times 10^{-3}$ mol dm^{-3}; $Q \ll K$) makes the dissociative pathway dominant, and a pseudo-first-order study under conditions where equilibrium lies far to the left ($[B] = 1.0 \times 10^{-3}$ mol dm^{-3}, $[D] = 0.30$ mol dm^{-3}; $Q \gg K$) makes the associative pathway dominant. Thus, incomplete studies of forward and reverse reactions could appear to give results inconsistent with the principle of microscopic reversibility. There is, of course, nothing in the complete rate law that is inconsistent with the principle of microscopic reversibility.

It is the concentrations of ligand species $B(CH_3CCPh)$ and $D(PPh_3)$ that determine the relative contributions of each pathway to the reaction, both forward and reverse, and it is the value of Q relative to K (0.70) that determines the direction of the net chemical change. Figure 14 presents some points about this rate law.

The rate law allows the writing of

$$R = 13.7 \text{ dm}^3 \text{ mol}^{-1} [D] + 0.055 \text{ dm}^3 \text{ mol}^{-1} [B]$$

for R, the ratio of the contributions to the reaction rate by the associative and dissociative pathways, and Figure 14a shows the lines for $R = 0.100$, 1.00, and 10.0. An equation can be written also relating the concentrations of ligand species to the value of Q/K and the ratio of the concentrations of platinum species $[A]/[C]$:

Figure 14. The relative importance of associative and dissociative pathways for the reversible reaction of tris(triphenylphosphine)platinum(0) with 1-phenyl-1-propyne.

$$R = \frac{\text{fraction by associative pathway}}{\text{fraction by dissociative pathway}}$$

Region at right, $R > 10$; region at left, $R < 0.1$; almost vertical line between these regions, $R = 1$.
(a) Each line at 45° corresponds to particular values of the product $(Q/K)([A]/[C])$
  Line 1 $(Q/K)([A]/[C]) = 0.10$
  Line 2 $(Q/K)([A]/[C]) = 1.00$
  Line 3 $(Q/K)([A]/[C]) = 10.0$
    Dashed line: $250[D] = [B]$, or denominator terms in second term of rate law are equal to one another.
    Points (a) $[B] = 0.30$ mol dm^{-3}, $[D] = 1.0 \times 10^{-3}$ mol dm^{-3}, $R = 0.030$; (b) $[B] = 1.0 \times 10^{-3}$ mol dm^{-3}, $[D] = 0.30$ mol dm^{-3}, $R = 4.1$; (c) $[B] = 0.10$ mol dm^{-3}, $[D] = 0.80$ mol dm^{-3}, $R = 11.0$.
(b) Each line at 45° corresponds to equilibrium ($Q/K = 1$) and indicated ratio $([A]/[C])_e$: 1. 10^{-3}; 2. 10^{-2}; 3. 10^{-1}; 4. 1; 5. 10^1; 6. 10^2; 7. 10^3.
  Lines originating at points (a) and (b) trace the approach to equilibrium:
  (a) $[A]_0 = [B]_0 = 0.30$ mol dm^{-3}, $[C]_0 = [D]_0 = 1.0 \times 10^{-3}$ mol dm^{-3} which goes to $[A]_e = [B]_e = 0.163$ mol dm^{-3}, $[C]_e = [D]_e = 0.138$ mol dm^{-3}
  (b) $[A]_0 = [B]_0 = 1.0 \times 10^{-3}$ mol dm^{-3}, $[C]_0 = [D]_0 = 0.30$ mol dm^{-3} which goes to $[A]_e = [B]_e = 0.163$ mol dm^{-3}, $[C]_e = [D]_e = 0.138$ mol dm^{-3}.

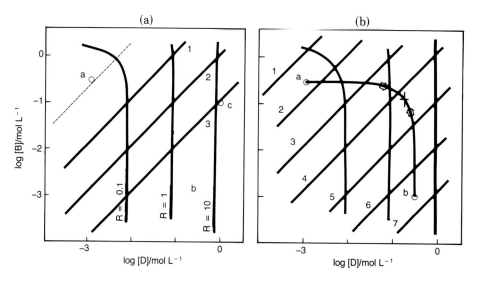

$$[D] = [B] \times \left(0.70 \frac{Q}{K}\right) \times \frac{[A]}{[C]}$$

The lines with slope 1 in this log–log plot correspond to particular values of the products of the quotients of Q/K and $[A]/[C]$. Another line of slope 1, the dashed line, is that which defines the relative concentrations of B and D at which the two denominator terms in the rate law for the dissociative pathway are equally important. This line corresponds to $250 [D] = [B]$. The two sets of experimental conditions described earlier are designated as points a and b in Figure 14a. The graph makes certain points clear. These are: (a) for the associative pathway to be dominant, the concentration of D, triphenylphosphine, must be high; this is a condition that tends to favor occurrence of the reverse reaction; (b) at low concentrations of D, the dissociative pathway is dominant regardless of the direction of the net reaction; and (c) extreme conditions are required for the associative pathway (the purely second-order term in the rate law) to be dominant for the forward reaction. (For the concentration conditions $[A] = 0.10$ mol dm^{-3}, $[B] = 0.10$ mol dm^{-3}, $[C] = 1.0 \times 10^{-3}$ mol dm^{-3}, and $[D] = 0.80$ mol dm^{-3}, $R = 10.9$, and $Q/K = 0.11$; these conditions are designated as point c in Figure 14a).

If a reaction is carried out under pseudo-first-order conditions with the total platinum(0) concentration very low compared with concentrations of the ligand species, with both ligands present initially, the entire course of a reaction from start to equilibrium is represented by a point on this graph. Thus points a and b in Figure 14a show the two experiments already described.

If two experiments like those described by points a and b in Figure 14a were run with the initial concentrations of reactants the same (0.300 mol dm^{-3}) and products the same (0.001 mol dm^{-3}), the course of these two reactions could be traced in Figure 14b, which shows that for each of the experiments the relative importance of the two pathways changes as reaction proceeds to equilibrium. The diagonal lines of slope 1 in this figure are drawn for $Q/K = 1$, i.e., equilibrium, and each line corresponds to a particular ratio of the concentrations of the two complexes at equilibrium.

References

(1) Thermodynamic data from JANAF Thermochemical Tables, Second Edition; Stull, D. R.; Prophet, H., project directors; NSRDS-NBS 37, 1971.
(2) Based in part upon King, E. L., Paper WSI 05, 7th Biennial Conference in Chemical Education, Stillwater, Okla., Aug. 1982.
(3) Cohen, R. W.; Whitmer, J. C. *J. Chem. Educ.* **1981**, *58*, 21.
(3a) Meyer, E. F. *J. Chem. Educ.* **1987**, *64*, 676.
(4) This type of graph presented in King, E. L. *Chemistry*; Painter-Hopkins: Sausalito, Calif., 1979.
(5) Atkins, P. W. *Physical Chemistry*, 2nd ed; W. H. Freeman: San Francisco, 1982; p. 149.
(6) Based in part upon King, E. L. *J. Chem. Educ.* **1981**, *58*, 975.
(7) Watson, I. D.; Williamson, A. G. *J. Chem. Educ.* **1979**, *56*, 723.
(8) May, D.; Rudd, D. F. *Chem. Eng. Sci.* **1979**, *31*, 59.
(9) Wentorf, R. H.; Hanneman, R. E. *Science* **1974**, *185*, 311; *Science* **1975**, *188*, 1037.
(10) Kelley, K. K.; King, E. G. *Bulletin 592*; U.S. Bureau of Mines, 1961.
(11) Brewer, L. *Chem. Rev.* **1953**, *52*, 1.
(12) Chao, R. E. *Ind. Eng. Chem. Prod. Res. Dev.* **1974**, *13*(2), 94.
(13) Abraham, B. M.; Schreiner, F. *Science* **1973**, *180*, 959; *Ind. Eng. Chem. Fundam.* **1974**, *13*, 305.
(14) Mesmer, R. E.; Baes, C. F., Jr. *Inorg. Chem.* **1969**, *8*, 618.
(15) Benson, S. W. *J. Am. Chem. Soc.* **1958**, *80*, 5151.
(16) van Holde, K. E. *Physical Biochemistry*; Prentice Hall: Englewood Cliffs, N.J., 1971; pp. 60–61.
(17) Pauling, L. *Proc. Natl. Acad. Sci. U.S.A.* **1935**, *21*, 186.
(18) Mills, C. C., III; King, E. L. *J. Am. Chem. Soc.* **1970**, *92*, 3017.
(19) Kury, J. W.; Paul, A. D.; Hepler, L. G.; Connick, R. E. *J. Am. Chem. Soc.* **1959**, *81*, 4185.
(20) Based in part upon King, E. L. *Inorg. Chem.* **1981**, *20*, 2350, and King, E. L., Paper Ma 1-6, Twenty-third International Conference on Coordination Chemistry, Boulder, Colo.; 1984.
(21) Bjerrum, J. *Metal Ammine Formation in Aqueous Solution*; P. Haase and Sons: Copenhagen, 1957; pp. 154, 162.
(22) Persson, H. *Acta Chem. Scand.* **1971**, *25*, 543.
(23) Sze, Y. K.; Irish, D. E. *J. Solution Chem.* **1978**, *7*, 395.
(24) Monod, J.; Wyman, J.; Changeux, J.-P. *J. Mol. Biol.* **1965**, *12*, 88.
(25) Newton, T. W. *J. Phys. Chem.* **1958**, *62*, 943.
(26) Sullivan, J. C. *J. Am. Chem. Soc.* **1962**, *84*, 4256.
(27) Awtrey, A. D.; Connick, R. E. *J. Am. Chem. Soc.* **1951**, *73*, 4546.
(28) Snyder, M. P.; Margerum, D. W. *Inorg. Chem.* **1982**, *21*, 2545.
(29) Ashurst, K. G.; Higginson, W.C.E. *J. Chem. Soc.* **1953**, 3044.
(30) Schwarz, H. A.; Comstock, D.; Yandell, J. K.; Dodson, R. W. *J. Phys. Chem.* **1974**, *78*, 488.
(31) de Leeuw, K. *Calculus*; Harcourt, Brace, and World: New York, 1966; pp. 160–62.
(32) Johnston, H. S. *J. Am. Chem. Soc.* **1951**, *73*, 4542.
(33) Based upon King, E. L. *J. Chem. Educ.* **1986**, *63*, 21.
(34) Wagman, D. D. et al. *J. Phys. Chem. Ref. Data* **1982**, *11*, Suppl 2, 38, 64–5.
(35) Nitrogen dioxide disproportionation: Denbigh, K. G.; Prince, A. J. *J. Chem. Soc.* **1947**, 790.
(36) Nitrous acid decomposition: Bray, W. C. *Chem. Rev.* **1932**, *10*, 161 reviews work of Abel, E.; Schmidt, H. *Z. Physik Chem.* **1928**, *132*, 56–64.
(37) Horiuti, J.; Nakamura, T. *Adv. Catalysis* **1967**, *17*, 47.
(38) Halpern, J.; Weil, T. A. *Chem. Comm.* **1973**, 631.

Chapter 9: PHYSICAL CHEMICAL ANALYSIS OF BIOPOLYMER SELF-ASSEMBLY INTERACTIONS

M. THOMAS RECORD, JR., AND BROUGH RICHEY
Departments of Chemistry and Biochemistry
University of Wisconsin—Madison
Madison, Wis. 53706

Understanding the thermodynamics and kinetics of noncovalent interactions between biopolymers in aqueous solution is fundamental to understanding the chemical basis of living systems. These subjects can be readily developed in an undergraduate course in macroscopic physical chemistry in the context of solution phase chemical thermodynamics, kinetics, and transport. However, solution phase physical chemistry is generally taught as a classical subject (if it is taught at all) in the modern undergraduate curriculum, and is therefore generally perceived as being of little relevance to the frontiers of chemistry. This is unfortunate, since the physical chemistry of processes and equilibria involving small or large molecules in solution provides the conceptual basis for analysis of biochemical processes, as well as being of general importance for the great majority of chemists involved in analytical, organic, polymer, and industrial work.

The physical chemistry of biopolymers and their complexes is an active research area. A long-standing focus of biophysical research has been determination of the structures of biopolymers and their assemblies, using X-ray diffraction and a wide range of other spectroscopic techniques. Molecular biology has evolved as a partnership between biophysical chemists and biologists; modern enzymology represents an analogous partnership between biophysical and bioorganic chemists. In both areas the traditional goal has been the development of structure–function relationships. Dramatic recent advances in biotechnology now make it possible to prepare and purify large quantities of site-specific sequence variants of proteins and nucleic acids. (No other area of polymer chemistry possesses the ability to prepare homogeneous samples of any desired molecular weight and monomer sequence.) This technological capability makes it possible in principle to modify or redesign existing proteins and nucleic acid sequences for biomedical or industrial purposes. Currently, however, few rules exist to relate protein or nucleic acid sequence to the formation of functional (three-dimensional) folded structures or complexes. Only in the case of the DNA double helix are the principles relating sequence to structure and function well-known, and even here the general relationships between DNA sequence and the regulation of expression of genetic information by specific proteins are as yet unclear. The current challenge for the biophysical chemist is to understand the relationship between structure and function at a fundamental thermodynamic and mechanistic level. What are the physical chemical principles underlying the thermodynamic stability, kinetics of formation, and lifetime of functional structures of biopolymers and their complexes? What are the physical chemical principles responsible for the specificity and the control of the processes these biopolymers perform? To answer these questions, one must begin by understanding the fundamental characteristics of noncovalent interactions in aqueous solution.

The limited purpose of this article is to illustrate the use of concepts from macroscopic solution phase physical chemistry to describe the general thermodynamic principles of noncovalent interactions of biological macromolecules. We begin with a brief description of cellular self-assembly processes and the types of noncovalent interactions that they utilize. We then present a thermodynamic analysis of the two principal driving forces for noncovalent associations involving nonpolar and highly charged solutes in aqueous solution, namely the hydrophobic effect and the polyelectrolyte effect. We discuss these topics at the level used in the upper level biophysical chemistry course at the University of Wisconsin—Madison, which is taught yearly to more than 150 seniors and graduate students in biochemistry, molecular biology, and chemistry. We conclude with several sample problems from this course that utilize the physical principles presented in these two examples of noncovalent assembly of biopolymers.

Noncovalent interactions in biological systems

At a physical chemical level, the fundamental characteristic of biopolymers (nucleic acids, proteins, carbohydrates, and lipids) is their capacity for *self-assembly* into ordered native structures. The organizational process of self-assembly is driven by the formation of *cooperative*, predominantly *noncovalent, interactions* and is specified by the primary chemical structure (sequence) of the biopolymers. The resultant native molecular and supramolecular assemblies have one or more of the following functions: definition of cellular form; self-replication, transfer, and processing of genetic information; catalysis of metabolic and biosynthetic chemical reactions; and coupling or transduction of free energy. Both the ability of cellular macromolecules to assemble and their function are regulated at the thermodynamic and kinetic levels by noncovalent interactions with effector ligands and other solution components, as well as by covalent modification and by physical variables (temperature, pressure).

Figure 1 is a block diagram summarizing the steps by which the components of the bacterium *E. coli* are synthesized and assembled from glucose and inorganic chemicals (1). We are concerned here with the assembly processes that follow synthesis of biopolymers of specific sequences. Self-assembly systems utilize one or more of the following noncovalent interactions in aqueous solution: association of nonpolar functional groups to reduce their interaction with water (the hydrophobic effect); association of oppositely charged oligo-ions or polyions to reduce their interaction with electrolyte ions (the polyelectrolyte effect); aromatic ring (π system) interactions; and hydrogen bonding. Since noncovalent interactions are individually weak, the *cooperative* formation of multiple interactions is invariably observed in the assembly of stable biopolymer structures or complexes.

To understand noncovalent interactions and their thermodynamic contributions to the stability of biomolecular assemblies, it is essential to appreciate the fact that in aqueous solution these types of interactions are *exchange reactions* that involve solutes and solvent species. Thus, energetic principles obtained from studies of noncovalent interactions in model systems in the gas phase (which lack this exchange component) are not directly applicable. The exchange character of the noncovalent interactions of solutes in water may be illustrated by use of a stoichiometric treatment of solvent and solute exchange in which fundamental nonideality effects are given an equivalent thermodynamic representation.

1. The hydrophobic effect. Nonpolar (hydrophobic) regions of solute molecules (R) associate to form a micelle or aggregate (R_m) and release water of hydrophobic hydration (2):

$mR(H_2O)_n \rightleftharpoons R_m + nm\ H_2O$

2. The polyelectrolyte effect. Two polyelectrolytes (P^{z+}, D^{z-}) interact in a univalent salt (MX) solution to form a complex (PD) and release associated counterions (3):

$P^{z+}(X^-)_n + D^{z-}(M^+)_m \rightleftharpoons PD + nX^- + mM^+$

3. Hydrogen bonding. A hydrogen bond donor/acceptor pair A and B forms a hydrogen bonded complex and releases water molecules that were hydrogen bonded to A and B:

$-AH \cdots OH_2 + -B \cdots HOH \rightleftharpoons -AH \cdots B- + HOH \cdots OH_2$

Figure 1. Covalent and noncovalent assembly reactions involved in the chemical synthesis of a bacterial cell from simple starting materials.

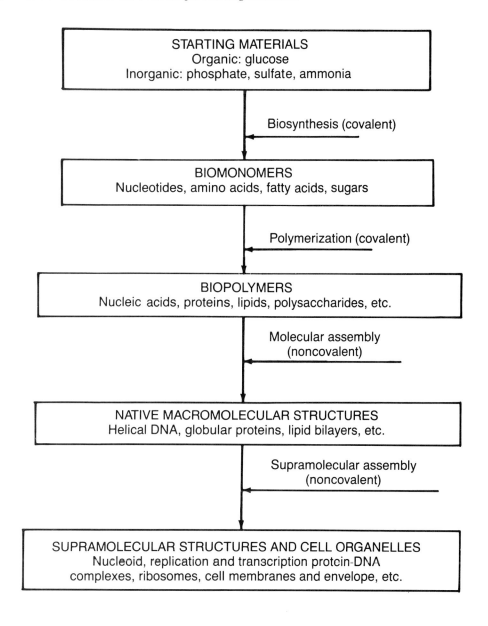

The stoichiometric representation in example 3 is a reasonable description of the hydrogen bond exchange reaction at both the molecular level and the thermodynamic level. On the other hand, the molecular pictures of hydrophobic hydration and counterion-polyelectrolyte interactions are much more complex than the simple stoichiometric formulations of processes 1 and 2 above would suggest. These interactions are not site-specific but diffuse, and are most appropriately described in terms of activity coefficients or thermodynamic interaction parameters. However, these nonideality effects can be incorporated into thermodynamically equivalent stoichiometric coefficients. This is analogous to the use of a thermodynamic activity in place of a concentration to account for nonideal mixing effects. In both cases, mathematical simplicity is retained although the physical picture is complicated.

The hydrophobic effect

Thermodynamic Description: The Process of Transferring a Nonpolar Solute (Benzene) to Water

The term *hydrophobic effect* refers to the characteristic thermodynamic behavior observed in studies of the solubility of nonpolar solutes (expressed as the mole fraction of solute in a saturated solution, X_2^{ss}) in water (4). From the measured solubility and its dependences on temperature and pressure, it is straightforward to determine all the thermodynamic functions for the process of transferring a mole of solute from the pure nonaqueous phase to its dilute solution standard state condition in water (designated $\Delta \bar{G}_{tr}^0$, $\Delta \bar{S}_{tr}^0$, etc.). These standard transfer functions are a direct source of information about the thermodynamic effects of solute–water interactions in the dilute solution environment, because they contain the composition-invariant activity coefficient describing the deviations from ideal mixing behavior that are due to interactions between the solute and water in the concentration range of the "ideal dilute" solution.

Figure 2 depicts the thermodynamic functions $\Delta \bar{G}_{tr}^0$, $\Delta \bar{H}_{tr}^0$ and $T\Delta \bar{S}_{tr}^0$ for the process of dissolving benzene in water as a function of temperature and includes the temperature dependence of the solubility ($\ln X_2^{ss}$) (5).

Figure 2. The thermodynamics of a hydrophobic effect, illustrated by the process of dissolving benzene in water. The figure is based on calorimetric measurements obtained over the temperature range 288–308 K (5) and was extrapolated beyond this range by assuming that $\Delta \bar{C}_{p,tr}^0$ is independent of temperature outside of this range.

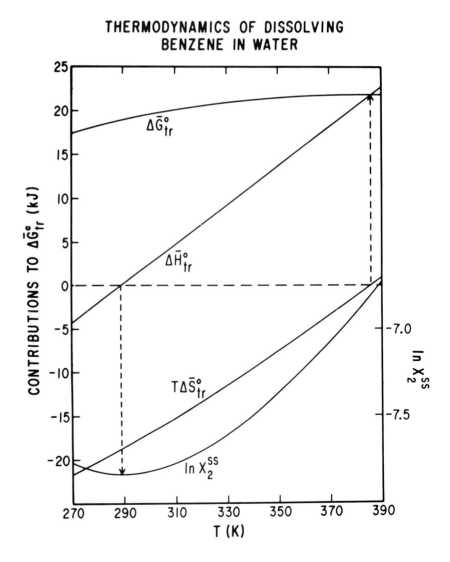

(This figure is based on calorimetric data for the temperature range 288–308 K, which are in good agreement with solubility data over the range 278–338 K [5]. Extrapolation beyond this range was carried out by assuming a constant $\Delta \bar{C}_{p,tr}^0$ as discussed below.) Analogous results have been obtained for a variety of hydrocarbons. Benzene is relatively insoluble at all temperatures, leading to a large positive $\Delta \bar{G}_{tr}^0$

when the mole fraction scale is employed. However, the large unfavorable value of $\Delta \bar{G}_{\text{tr}}^0$ cannot be uniquely attributed to a dominant enthalpic or entropic term, since both $\Delta \bar{H}_{\text{tr}}^0$ and $\Delta \bar{S}_{\text{tr}}^0$ are strongly temperature dependent and indeed change sign in the temperature range shown. These strong temperature dependences result from a large positive $\Delta \bar{C}_{p,\text{tr}}^0$, which has its origin in the anomalously large standard partial molar heat capacity observed for benzene in water. *Large, temperature-insensitive values of $\Delta \bar{C}_{p,\text{tr}}^0$ are in fact the principal thermodynamic characteristic of processes that expose nonpolar functional groups to water.* The fact that $|\Delta \bar{C}_{p,\text{tr}}^0| \gg |\Delta \bar{S}_{\text{tr}}^0|$ over a wide temperature range gives rise to the general thermodynamic phenomenon of entropy–enthalpy compensation, whereby both $\Delta \bar{H}_{\text{tr}}^0$ and $T\Delta \bar{S}_{\text{tr}}^0$ change much more rapidly with temperature than does $\Delta \bar{G}_{\text{tr}}^0$ (see Figure 2). Entropy–enthalpy compensation is a general characteristic of processes in which the degree of exposure of nonpolar groups to water is affected.

For many nonpolar solutes (including benzene), a temperature of minimum solubility in water is observed near 20 °C. At this temperature, $\Delta \bar{H}_{\text{tr}}^0 = 0$, and the process of dissolving benzene is unfavorable for purely entropic reasons. (At lower temperatures, $\Delta \bar{H}_{\text{tr}}^0 < 0$ so that the enthalpic term is favorable; the standard process remains unfavorable in this temperature range because of the dominant unfavorable entropic term.) As the temperature is increased above that of minimum solubility, the entropic term, though still unfavorable, becomes less dominant; concurrently the enthalpic term becomes more unfavorable. If the extrapolation in Figure 2 is assumed to be valid, a temperature is eventually reached where $\Delta \bar{S}_{\text{tr}}^0$ passes through 0, at which point $\Delta \bar{G}_{\text{tr}}^0$ is a maximum and the transfer process becomes unfavorable for purely enthalpic reasons. (Numerous calculations involving temperature derivatives or integrals of the transfer functions may be used to illustrate aspects of the behavior of this model hydrophobic system. One example, involving the thermodynamic effects of transferring a nonpolar functional group of a solute from H_2O to D_2O, is given as Problem 1 in the last section of this chapter.)

The unique thermodynamic characteristics of the standard transfer process in which nonpolar solutes are exposed to water may be interpreted in terms of a hydrophobic hydration layer $(R[H_2O]_n)$ (2). Unlike the concept of hydration of ions or polar solutes, however, the concept of a hydrophobic hydration layer refers to the thermodynamic consequences of a diffuse "ordering" of water in the vicinity of the solute, and as such does not involve a large favorable enthalpy of interaction with the solute. The large $\Delta \bar{C}_{p,\text{tr}}^0$ for this process indicates that the water of hydrophobic hydration is highly thermolabile. In other words, the average number of water molecules bound (n) must decrease rapidly with increasing temperature. (An oversimplified but didactically useful explanation of the large $\Delta \bar{C}_{p,\text{tr}}^0$ is that there is an equilibrium of the form $R(H_2O)_n \rightleftarrows R + nH_2O$ that shifts to the right with increasing temperature [2].)

In view of the strong temperature dependences of $\Delta \bar{H}_{\text{tr}}^0$ and $\Delta \bar{S}_{\text{tr}}^0$ (and the fact that these quantities change sign at different temperatures for different nonpolar solutes), it is not surprising that no general correlations have been found between the values of $\Delta \bar{H}_{\text{tr}}^0$ or $\Delta \bar{S}_{\text{tr}}^0$ and the size or shape of the nonpolar solute. On the other hand, since the values $\Delta \bar{G}_{\text{tr}}^0$ and $\Delta \bar{C}_{p,\text{tr}}^0$ are relatively insensitive to changes in temperature, reasonable correlations exist that relate these quantities to the surface area (or other molecular properties) of the nonpolar solute. In the case of $\Delta \bar{G}_{\text{tr}}^0$, it is found that exposure of 1 Å^2 of nonpolar surface to water is accompanied by an increase in the unfavorable standard free energy of transfer of approximately 80–100 J/mol near 20 °C (6). Similarly, the value of $\Delta \bar{C}_{p,\text{tr}}^0$ increases by about 1.4 J/mol K per additional Å^2 of nonpolar surface exposed to water (7).

The solubility of nonpolar solutes increases with increasing pressure at constant temperature, indicating that the standard volume change of transfer, $\Delta \bar{V}_{\text{tr}}^0$, is negative. For benzene, $\Delta \bar{V}_{\text{tr}}^0 = -6.2$ cm^3/mol; values near -20 cm^3/mol are obtained for various small alkanes (8). The sign of the volume change indicates that hydrophobic hydration does not resemble the formation of an open ice-like lattice of water in the vicinity of the nonpolar solute. Whatever the correct molecular picture, processes exposing nonpolar groups to water are favored by an increase in pressure because of the decrease in the standard partial molar volume of the solute that occurs upon transfer to water.

The Contribution of the Hydrophobic Effect to the Stability of Globular Proteins in Aqueous Solution

Contributions of the various noncovalent interactions to the stability of the native state of globular proteins in aqueous solution have been investigated by differential scanning microcalorimetry (9–11). The excess heat capacity of a protein solution (relative to solvent) is measured as a function of temperature in the vicinity of the cooperative conformational transition between the native folded structure and the unfolded, flexibly coiling chain molecule. These measurements provide a relatively complete thermodynamic description of the transition in terms of the following quantities: (1) the transition midpoint temperature (T_m); (2) the (integrated) calorimetric molar transition enthalpy ($\Delta \bar{H}^0$); (3) the temperature dependence of the apparent equilibrium constant for the two-state model of the transition (native $\leftrightarrows$ denatured), (K_{eq}); (4) the apparent van't Hoff transition enthalpy (ΔH^0_{vH}), evaluated from the temperature dependence of K_{eq}; and (5) the change in partial molar heat capacity accompanying denaturation of the protein ($\Delta \bar{C}^0_p$), as evaluated from the difference between the high- and low-temperature calorimetric baselines. For small globular proteins with a single domain, the calorimetric enthalpy $\Delta \bar{H}^0$ and the van't Hoff enthalpy ΔH^0_{vH} agree to within 5%. This demonstrates that the denaturation reaction for these proteins is (to a good approximation) a completely cooperative process at the thermodynamic level and that no stable intermediate species exist at significant levels at equilibrium in the transition region.

Figure 3 summarizes the thermodynamic analysis of the stability of myoglobin (Mb), a small oxygen-binding protein found in vertebrates. The figure is based on data (obtained by differential scanning microcalorimetry) that show that $\Delta \bar{C}^0_p$ for the thermal denaturation process is independent of temperature (9). The thermodynamics of the process of denaturing Mb are dominated by a large $\Delta \bar{C}^0_p$ ($\Delta \bar{C}^0_p$ = 11.6 kJ K^{-1}mol^{-1}) and the accompanying entropy/enthalpy compensation phenomenon. Clearly the thermodynamic description of the process of denaturing Mb is parallel to that of the process of transferring benzene to water discussed above.

For Mb denaturation at its transition temperature T_m, $K_{eq} = 1$, $\Delta \bar{G}^0 = 0$, and $\Delta \bar{H}^0 = T_m \Delta \bar{S}^0$, as for a phase equilibrium in a pure system. Below T_m, the globular native state is stable and both $\Delta \bar{H}^0$ and $T\Delta \bar{S}^0$ decrease strongly with decreasing temperature, resulting in a very gradual increase in stability with decreasing temperature. The stability of the native state reflects a delicate balance between generally large and opposing contributions from enthalpic and entropic factors. For Mb and the other single-domain proteins investigated, the native state is stable by no more than 40–60 kJ mol^{-1} (i.e., a few hundred Joules per mole of amino acid residues) in the physiological temperature range (9). Myoglobin and a few other proteins exhibit an experimentally accessible temperature of maximum stability (T_s) above 0 °C. At T_s, $\Delta \bar{G}^0$ is a maximum, $\Delta \bar{S}^0 = 0$, and $\Delta \bar{G}^0 = \Delta \bar{H}^0$. (As shown in Figure 3, K_{eq} exhibits a minimum at a somewhat lower temperature than T_s; the minimum in K_{eq} occurs at the temperature at which $\Delta \bar{H}^0 = 0$.) Further extrapolation of the thermodynamic behavior to temperatures lower than T_s indicates that the stability of Mb decreases with decreasing temperature in this range, leading to the prediction of a second denaturation transition at low temperature. Privalov and co-workers have very recently demonstrated the existence of this "cold-denaturation" transition of myoglobin (10). (Calculations of the thermodynamic features of this stability diagram from the calorimetric data make excellent problem assignments, especially if a temperature-independent $\Delta \bar{C}^0_p$ is assumed. Problem 2 at the end of this chapter provides one such example, in which the Clapeyron equation for dT_m/dP is used to analyze the phase diagram of stability of a typical globular protein.)

The Contribution of the Hydrophobic Effect to Other Self-Assembly Processes

Micelle formation from amphipathic monomers and aggregation of cytoskeletal proteins (e.g., tubulin, actin, myosin) are cooperative intermolecular self-assembly processes whose thermodynamic behavior is directly analogous to that observed in the cooperative intramolecular self-assembly (folding) of a protein chain. The critical

Figure 3. The thermodynamics of myoglobin denaturation, illustrating the importance of the hydrophobic effect on protein stability. The figure is based on calorimetric data which show that the value of $\Delta \bar{C}_p^0$ for the denaturation reaction is independent of temperature over the range 323–353 K (9, 10). Values of the thermodynamic functions outside this range were calculated by assuming $\Delta \bar{C}_p^0$ is constant over the entire temperature range shown.

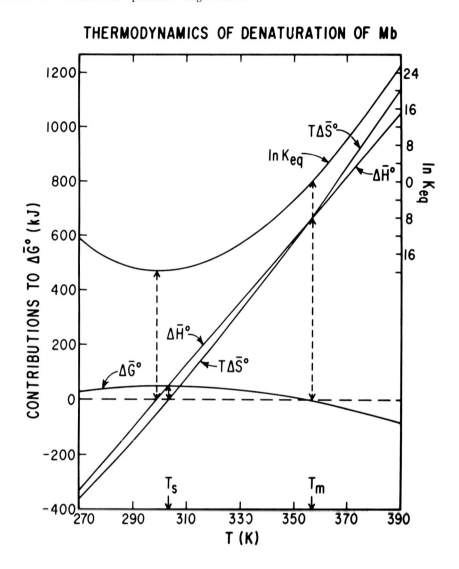

monomer concentration X_c is the fundamental thermodynamic variable in these intermolecular assembly processes. This quantity, which is the analogue of a solubility, represents the concentration of free monomer in equilibrium with the micelle or aggregated species. In situations where the number of monomers in the micelle or aggregate is large,

N monomer $\leftrightarrows$ aggregate $\qquad N \gtrsim 20$

the aggregate behaves thermodynamically like a microphase, whose entropy of mixing with solvent may be neglected. Consequently, for large aggregates, the aggregation equilibrium is equivalent to a solubility equilibrium, and the critical monomer concentration plays the role of a solubility. (Consideration of equilibria of the form $NA \leftrightarrows A_N$ is a valuable teaching tool to help bridge the gap between noncooperative chemical equilibria [$N = 2$] and completely cooperative phase equilibria [$N = \infty$]. Aggregation and micelle formation are examples of an intermediate situation. Both processes become increasingly analogous to phase equilibria as N increases).

From X_c and its dependences on temperature and pressure, apparent thermodynamic functions (neglecting monomer nonideality) for the process of transferring a monomer from an aqueous environment to the interior of a micelle or aggregate can be ascertained. These processes typically exhibit the thermodynamic behavior expected for the removal of nonpolar groups from an aqueous environment: $\Delta \bar{C}_p^0$ is large in magnitude and negative in sign, enthalpy–entropy compensation is observed, and a minimum in X_c often occurs in or near the physiological temperature range. An example of such behavior for myosin (12) is included as Problem 3 later in this chapter.

Equilibrium constants for the binding of nonpolar ligands to hydrophobic regions of proteins also show the characteristic dependences on temperature and pressure expected for processes driven by the release of hydrophobically associated water. Similar dependences are observed for binding reactions, which couple ligand binding to conformational changes that alter the amount of biopolymer hydrophobic surface exposed to water. Problem 4 at the end of this chapter explores these effects in more detail. In this example, the hydrophobic component of the interaction of the enzyme RNA polymerase with a specific (promoter) site on DNA is attributed to a conformational change in the polymerase that reduces the exposure of nonpolar surface to water (13). Other examples involving the hydrophobic contribution to ligand binding and conformational change include the binding of octanol to bovine serum albumin (14), the specific and nonspecific binding of tRNA to a tRNA synthetase (15), the binding of avidin to biotin (16), and the partitioning of hydrocarbons between water and dodecylsulfate micelles (17).

The polyelectrolyte effect

The thermodynamics of ion–polyelectrolyte interactions are best taught in the context of the Donnan membrane equilibrium. Such an approach takes advantage of the students' background in membrane equilibria involving nonelectrolytes (osmotic pressure, equilibrium dialysis) and also provides a point of departure for discussing electrochemical potentials, membrane potentials, and chemiosmotic coupling in cellular bioenergetics. Though a classical subject, the Donnan membrane equilibrium is at the cutting edge of current theoretical research on polyelectrolyte solutions, since the cell model that is used in most statistical mechanical analyses (Poisson–Boltzmann theory, Monte Carlo calculations) is conceptually equivalent to a microscopic Donnan compartment. In fact, grand canonical Monte Carlo cell model simulations of the ion distributions surrounding polyelectrolytes are a close theoretical analogue of an actual microscopic Donnan equilibrium experiment, since they calculate the number of ions contained within the cell at constant chemical potential of the electrolyte component (18).

A didactically useful means of presenting the principles of the Donnan effect is to consider the redistribution of electrolyte ions that follows the addition of a membrane-impermeable polyelectrolyte to one compartment (α) of a cell containing an aqueous salt solution and a membrane partition that is permeable to both water and electrolyte ions. This situation is illustrated schematically in Figure 4 for the case of a polyanion (NaDNA) and a univalent electrolyte (NaCl). In Figure 4, the unknown x represents the change in concentration of the electrolyte component (or equivalently the change in concentration of the Cl$^-$ co-ion) that occurs upon establishing membrane equilibrium after the addition of NaDNA. A key conceptual question at this point is why the co-ion concentration should be affected at all by addition of the electroneutral polyelectrolyte component. No net transfer of co-ions need occur for reasons of electroneutrality. Electroneutrality is preserved for all values of x, including $x = 0$. (Electroneutrality requires only that the numbers of Na$^+$ and Cl$^-$ ions transferred from phase α to phase β be identical.) The driving force for co-ion redistribution is an entropic (mixing) effect. In the absence of ion redistribution ($x = 0$), there is a large transmembrane concentration gradient of Na$^+$ and no transmembrane concentration gradient of Cl$^-$. A more random and therefore entropically more favorable situation is obtained by shifting NaCl from phase α to phase β. This results in a more equal distribution of particles across the membrane and in smaller, opposing transmembrane concentration gradients of both ions. If random mixing were the only factor driving reequilibration, the equilibrium value of x for NaCl would be 0.25 m$_2$ (implying the transfer of one quarter of a mole of NaCl per mole of added DNA monomer). For $x = 0.25$ m$_2$, the differences in concentration

Figure 4. The Donnan membrane experiment. The addition of m_2 mol of NaDNA to side α of a Donnan cell causes x mol of NaCl to be transferred across the semipermeable membrane to side β.

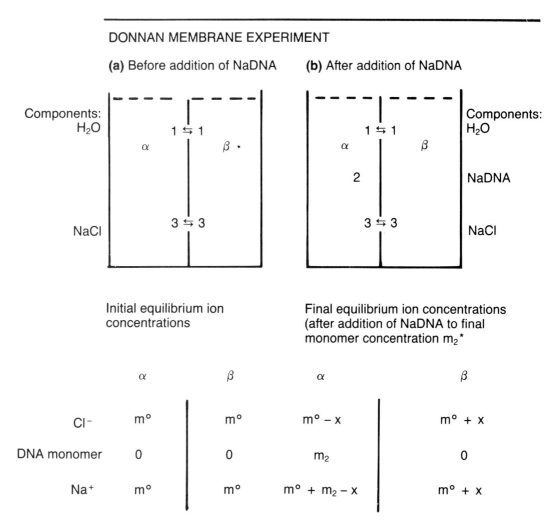

of Na$^+$ and of Cl$^-$ across the membrane are equal in magnitude but opposite in sign ($m_{Cl,\alpha} - m_{Cl,\beta} = -0.5\,m_2$; $m_{Na,\alpha} - m_{Na,\beta} = 0.5\,m_2$). This case is appropriately referred to as the *ideal Donnan distribution* and is readily obtained from the fundamental condition for membrane equilibrium either in terms of electrolyte chemical potentials ($\mu_{3,\alpha}^{eq} = \mu_{3,\beta}^{eq}$), or in terms of single-ion electrochemical potentials. In either analysis of the ideal case, osmotic pressure effects are neglected and it is assumed that the mean ionic activity coefficient of the electrolyte ($\gamma_\pm$) is the same on both sides of the membrane at equilibrium. The latter assumption is equivalent to neglecting ion–polyion interactions and can be shown to be *totally inappropriate* for all but the most weakly charged polyelectrolytes. Nevertheless the ideal Donnan distribution is a conceptually useful reference state for comparison with actual Donnan distributions observed for highly charged polyelectrolytes.

As in the general analysis of equilibrium dialysis, the equilibrium Donnan ion distributions are characterized by experimentally accessible distribution coefficients Γ_{Cl} and Γ_{Na}:

$$\Gamma_{Cl} \equiv \frac{m_{Cl,\alpha} - m_{Cl,\beta}}{m_2} \qquad \Gamma_{Na} \equiv \frac{m_{Na,\alpha} - m_{Na,\beta}}{m_2}$$

The thermodynamic distribution coefficients are generally interpreted in terms of site-binding models in equilibrium dialysis studies of ligand–polymer interactions. However, the molecular interpretation of Donnan ion distribution coefficients is more complex than this. For the ideal Donnan distribution, $\Gamma_{Cl} = -0.5$ (co-ion exclusion) and $\Gamma_{Na} = 0.5$ (counterion accumulation). The limiting value of the co-ion distribution coefficient obtained as m_2 approaches 0 is called the Donnan coefficient (Γ_{Cl}^0), which is equivalent to the limiting value of the thermodynamic interaction parameter Γ_{32}^0 (19):

$$\Gamma_{Cl}^0 \equiv \lim_{m_2 \to 0} \frac{m_{Cl,\alpha} - m_{Cl,\beta}}{m_2} = \lim_{\Delta m_2 \to 0} \frac{\Delta m_3}{\Delta m_2} = \left(\frac{\partial m_3}{\partial m_2}\right)_{\mu_1,\mu_3,T} \equiv \Gamma_{32}^0$$

Here the symbol Δ refers to the difference between side α and side β and the condition of constant μ_1, μ_3, and T corresponds to the thermodynamic condition of Donnan membrane equilibrium.

Donnan distribution coefficients have been determined for DNA as a function of electrolyte concentration (20). Closed-form expressions for Γ_{Cl}^0 and Γ_{32}^0 have been derived from the Poisson–Boltzmann cylindrical cell model (21, 22) and from counterion condensation theory (23). Numerical values of these quantities have been determined by grand canonical Monte Carlo simulations of models of DNA in solution (18). Under the so-called limiting-law conditions of low but excess uni-univalent salt, the various theories predict that Γ_{Cl}^0 and Γ_{32}^0 are a function only of the dimensionless axial charge density parameter $\xi \equiv e^2/\epsilon kTb$. (This Bjerrum-like parameter is the ratio of the Coulombic energy to the thermal energy for the interaction of adjacent fixed charges on the polyion that are separated by a distance b in a medium of dielectric constant ϵ.) Helical DNA is a polyanion of high axial charge density (two phosphate groups per 3.4 Å, $\xi = 4.2$), and under limiting-law conditions all theories predict that in a solution of helical DNA, $\Gamma_{Cl}^0 = -(4\xi)^{-1} = -0.06$. This theoretical result is in reasonable agreement with experiment. (At the polyion and electrolyte concentrations at which these membrane equilibria are studied, the effect of polyion excluded volume on the Donnan ion distribution coefficients complicates the comparison of the experimental data with theoretical limiting-law values [24].)

The difference between the ideal and limiting-law values of Γ_{Cl}^0 for DNA illustrates the *overwhelming* importance of nonideality in determining the thermodynamic properties of solutions of highly charged polyelectrolytes. Contrary to the behavior of ordinary ionic solutions, polyelectrolyte nonideality is most important at low salt concentrations. The ion distribution surrounding a highly charged polyion becomes increasingly nonuniform (nonrandom) as the salt concentration is reduced. Both Poisson–Boltzmann and Monte Carlo calculations predict that, for helical DNA, Γ_{Cl}^0 decreases from the nonideal limiting-law value (-0.06) toward the ideal value (-0.50) as the salt concentration is increased.

The effect of electrolyte–polyion interactions may be appreciated by comparing the ideal and actual extents of redistribution of NaCl in the approach to equilibrium following addition of NaDNA in a membrane experiment (cf. Figure 4). The actual NaCl concentration shift in the approach to equilibrium under limiting-law conditions is $x = 0.03\,m_2$. The ideal value is $x = 0.25\,m_2$. Much less NaCl is transferred across the membrane than is predicted by the ideal calculations, and thus the Na$^+$ concentration difference across the membrane is much larger than the ideal value ($\Gamma_{Na}^0 = 0.94$ under limiting-law conditions, as compared with $\Gamma_{Na}^0\,\text{ideal} = 0.50$). In other words, the DNA behaves as if it retains most of its Na$^+$ counterions; random mixing of ions is not achieved.

The macroscopic distribution coefficients discussed above accurately reflect the local ion distributions in the vicinity of the individual polyanions; i.e., there is a local exclusion of Cl$^-$ and accumulation of Na$^+$ in the vicinity of a DNA molecule. For helical DNA, the local concentration of Na$^+$ near the DNA surface is predicted to be in the *molar* range. The polyion axial charge density (ξ) is the primary factor that determines the local counterion concentration at the DNA surface. Perhaps surprisingly, changes in the bulk salt concentration do not strongly affect the local counterion concentrations at the DNA surface. The locally high counterion concentration is effectively buffered against change by the polyion electric field and is preserved even in the absence of added salt at infinite dilution of the polyion. This apparent contradiction of the principle of mass action (which would predict entropy-driven counterion dissociation at high dilution) results from the prohibitive increase in electrostatic free energy that would arise from intramolecular repulsions among the fixed polyion charges if counterions dissociated from the local region surrounding the polyion.

The molecular picture of an NaDNA/NaCl solution that emerges from the above considerations is one in which much of the DNA structural charge is effectively neutralized by the accumulation of a high local concentration of counterions (Na^+). (Problem 5 at the end of this chapter leads the student through the Donnan analysis of Mg^{2+}–DNA interactions, and the ion exchange competition between Na^+ and Mg^{2+} for the vicinity of DNA.) Local ion gradients in the vicinity of a polyion are extreme examples of the Debye Hückel ion atmosphere formed about each ion in an ordinary salt solution. The extent of neutralization of polyion charge by accumulated counterions is found to be relatively independent of the bulk salt concentration and is determined primarily by the axial charge density of the polyion. The extreme degree of local counterion accumulation reduces the requirement for local co-ion (Cl^-) exclusion.

Although the large gradients in the concentrations of counterions and co-ions about the DNA polyanion are continuous, both the molecular properties and thermodynamic consequences of these gradients can be closely approximated by the simple two-state model of counterion condensation (23). In this model, the local region is considered to be a co-ion-free phase of uniformly high counterion concentration that effectively neutralizes much of the polyion charge, and thereby greatly reduces the ion gradients that extend into the bulk solution (23, 25). An even simpler picture is sufficient for thermodynamic purposes. The neutralization of polyion charge by the local ion concentration gradients is in fact thermodynamically equivalent to a stoichiometric degree of counterion binding (22, 26). In this thermodynamic description, highly charged polyelectrolytes are considered to be a special class of weak electrolytes for which the extent of counterion dissociation is essentially independent of the bulk salt concentration, even at extreme dilution. The counterion stoichiometric coefficient *completely accounts* for nonideality due to electrolyte–polyion interactions, while preserving the simple functional forms derived for the thermodynamic properties of ideal polyelectrolyte solutions.

For double helical NaDNA in NaCl, the limiting-law Donnan coefficient ($\Gamma^0 = -0.06$) can be interpreted stoichiometrically as resulting from the association of $1 + 2\Gamma^0 = 0.88$ Na^+ ions per DNA phosphate. In this thermodynamic interpretation, which has been rigorously justified (22), the total extent of Na^+ accumulation per polyion charge ($\Gamma^0 = 0.94$) is attributed to a sum of contributions from counterion association (0.88) and an ideal Donnan accumulation (0.06), whereas co-ion exclusion is attributed solely to the ideal Donnan exclusion (-0.06). We refer to the thermodynamic consequences of electrolyte–polyelectrolyte interactions as the *polyelectrolyte effect*, by analogy with the hydrophobic effect, a term that refers to the thermodynamic consequences of the interactions of nonpolar solutes with water (27, 28).

The use of the thermodynamic degree of counterion association provides a simple route to the analysis of electrolyte effects on noncovalent processes that alter the charge density of nucleic acids (helix–coil transitions, binding of oligocations and proteins) (22, 26, 29). For example, the cooperative conformational transition from the native double helix to two unstructured polymer chains is accompanied by a reduction in the structural (axial) charge density from two phosphates per 3.4 Å ($\xi = 4.2$) to one phosphate per ~4.1 Å ($\xi \simeq 1.7$ at pH 7). This corresponds to a decrease in the thermodynamic degree of counterion association from 0.88 Na^+ to 0.71 Na^+ per DNA phosphate. The transition process at the level of a DNA monomer (phosphate) is therefore summarized by the following stoichiometric equation:

Native DNA ($Na_{0.88}$DNAP) → Denatured DNA ($Na_{0.71}$DNAP) + 0.17 Na^+

Consequently, the denaturation process is favored by a reduction in the bulk salt concentration. This polyelectrolyte effect is observed experimentally as a 15–20 °C reduction in the transition temperature T_m per tenfold reduction in the bulk NaCl concentration.

As a second example, the binding of an oligocation L^{z+} to helical DNA may be represented stoichiometrically by the approximate ion exchange equation (neglecting possible anion association with L^{z+}):

$$L^{z+} + \text{DNA site } (\text{Na}_{0.88}\text{DNAP})_z \rightarrow \text{complex} + 0.88z\ \text{Na}^+$$

Consequently, the extent of complex formation varies with the salt concentration as $[\text{Na}^+]^{-0.88z}$. The specific and nonspecific interactions of proteins with DNA, which are the molecular basis for the control of gene expression, also behave as ion exchange reactions at the thermodynamic level. The polyelectrolyte effect (ion exchange stoichiometry) exhibited in these systems is typically so large that the extent of complex formation is *far* more sensitive to the salt concentration than to the protein concentration itself (27–29).

Representative problems treating the thermodynamics of noncovalent interactions

Problem 1. Transfer of a Nonpolar Amino Acid Side Chain from H₂O to D₂O: The Relative Strengths of the Hydrophobic Effect in H₂O and D₂O (30).

Relative solubilities, β (expressed as ratios of mole fractions in the saturated solutions), of the amino acids glycine($^+\text{H}_3\text{NCH}_2\text{CO}_2^-$) and phenylalanine ($^+\text{H}_3\text{NCHCO}_2^-$) have been determined in D₂O and H₂O at temperatures near 25 °C.
$\quad\quad\quad\ \ \ |$
$\quad\quad\quad\text{CH}_2(\text{C}_6\text{H}_5)$

		glycine	phenylalanine
$\beta \equiv X_{\text{D}_2\text{O}}^{ss}/X_{\text{H}_2\text{O}}^{ss}$	β	1.049	0.854
	$\dfrac{d\beta}{dT}$	-1.093×10^{-3} deg^{-1}	-1.489×10^{-3} deg^{-1}

Questions

1. Derive general expressions for the standard free energy, enthalpy, and entropy of transfer of a solute from H₂O to D₂O in terms of β and $d\beta/dT$. Start with expressions for the solute chemical potentials in the two solvents. Assume that solute–solute interactions are similar in H₂O and D₂O, so that for values of β near 1, the activity coefficients γ_2 reflecting deviations from ideal dilute solution behavior will be equal in the two solvents.

2. Calculate the standard thermodynamic functions (ΔG^0, ΔH^0, and ΔS^0) for the transfer of the phenylalanine side chain ($-\text{CH}_2\text{C}_6\text{H}_5$) from H₂O to D₂O at 25 °C and 1 atm.

3. Does the hydrophobic effect appear to be more pronounced or less pronounced in D₂O than in H₂O? Explain your reasoning briefly. What further experimental information would be helpful to strengthen this conclusion?

Problem 2. Thermal Denaturation of a Globular Protein

Denaturation of globular proteins in aqueous solution may be modeled as a phase change in a pure system, because the process is highly cooperative (two-state)

$$\text{N(native)} \rightarrow \text{D(denatured)}$$

and differential mixing effects are absent. For a particular protein, ΔH^0 of denaturation is 160 kcal/mol at the transition temperature (T_m) of 355 K and 1 atm. For this transition, $\Delta C_p^0 = 2.80$ kcal/mol deg and $\Delta V^0 = -20$ cm^3/mol. Assume both ΔC_p^0 and ΔV^0 are independent of temperature and pressure.

Questions

1. Derive expressions for the standard functions ΔH^0 and ΔS^0 as a function of temperature for the process of converting N to D at 1 atm pressure.

2. Derive expressions for the standard functions ΔH^0 and ΔS^0 as a function of pressure for the process of converting N to D at constant temperature.

3. Use your results from the above parts (and general reasoning) to explain quantitatively the features of the phase diagram for the stability of the N state of this protein (see Figure 5). Start with the general thermodynamic expression for the variation of T_m with P.

Figure 5. Phase diagram of the stability of a typical globular protein.

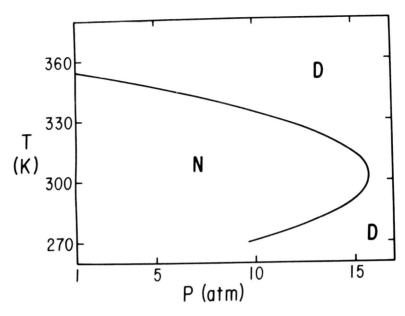

Problem 3. Aggregation of Myosin

The critical concentration for aggregation of the rodlike protein myosin in water at 20 °C, pH 7, and physiological salt concentration is 3.7×10^{-7} M. This critical concentration is independent of temperature near 20 °C.

Questions

1. Estimate $\Delta \bar{G}^0$, $\Delta \bar{H}^0$, and $\Delta \bar{S}^0$ for the aggregation process at 20 °C. State the approximation(s) required for this estimate. Explain the meaning and sign of ΔG^0 briefly.

2. Calculate $\Delta \bar{G}$ for myosin aggregation at 20 °C, pH 7, and a monomer concentration of 3.7×10^{-8} M. How much chemical work (per mole) would be required to drive myosin aggregation at this monomer concentration?

3. Protons are one source of chemical work to drive myosin aggregation. At pH 6 and 20 °C the critical concentration is in fact shifted to 3.7×10^{-8} M. Estimate the number of protons taken up or released per myosin monomer incorporated into an aggregate.

Problem 4. RNA Polymerase–DNA Interactions (*13, 27*)

The noncovalent interaction between the enzyme RNA polymerase (a large globular protein) and a specific binding site on DNA, called a promoter (P), may be summarized as $R + P \rightleftarrows RP_0$ where RP_0 is a complex in which the DNA has been locally opened so that polymerase can read the sequence of the strand to be transcribed. An equilibrium constant for this interaction may be defined as

$$K_{eq} = \frac{[RP_0]_{eq}}{[R]_{eq}[P]_{eq}}$$

K_{eq} is found to obey a temperature dependence of the form:

$$\ln K_{eq} = A_1 + A_2 \ln T + \frac{A_3}{T} \text{ where } A_1, A_2, \text{ and } A_3 \text{ are constants, and } T \text{ is in } K.$$

Questions

1. Derive general expressions for the standard thermodynamic functions ΔG^0, ΔH^0, ΔS^0, and ΔC_p^0 for the process of converting R and P to RP_0 at the temperature T under standard conditions. (Express your answers in terms of A_1, A_2, A_3, and T.)

2. For the interaction of RNA polymerase with the λP_R promoter under roughly physiological solution conditions:

$$A_1 = 8182.6 \qquad A_2 = -1207.8 \qquad A_3 = -380825$$

such that $K_{eq} = 1.2 \times 10^{11}$ M^{-1} at 37 °C. Evaluate the above standard thermodynamic functions for the process of converting R and P to RP$_0$ under standard conditions at 37 °C. Explain carefully the nature of the process for which ΔG^0 is the free energy change.

3. To what extent are your results for 2 above consistent with a hydrophobically driven process? Explain briefly.

4. Would you expect K_{eq} to increase or decrease with increasing pressure? Justify your choice with an equation and an explanation.

5. At low temperatures, RNA polymerase binds to this promoter to form a closed complex (RP$_c$), which undergoes a conformational change to form the open complex (RP$_0$).

$$RP_c \overset{K_0}{\rightleftharpoons} RP_0 \qquad K_0 = \frac{[RP_0]_{eq}}{[RP_c]_{eq}}$$

At 0 °C, $K_0 = 0.025$, and this promoter is almost completely closed. What expenditure of chemical work would be required at 0 °C to open this promoter and achieve a condition in which [RP$_0$]/[RP$_c$] = 25?

Problem 5. Mg^{++}–DNA Interactions

Mg^{2+} interacts with DNA and other highly charged polyanions to an extent that is quantitatively different than that of monovalent cations like Na$^+$. This problem explores Mg^{2+}–DNA interactions via membrane equilibrium dialysis. Recall that for the general strong electrolyte (MA $\to \nu_+ M^{z+} + \nu_- A^{z-}$)

$$a_{MA} \equiv a_M^{\nu_+} a_A^{\nu_-} = (\gamma_\pm)^{\nu_+ + \nu_-} m_M^{\nu_+} m_A^{\nu_-}$$

Questions

1. For an aqueous MgCl$_2$ solution, write a general expression for the activity of the salt (a_{MgCl_2}) in terms of the mean ionic activity coefficient ($\gamma_\pm$) and the molal concentrations of the individual ions ($m_{Mg^{2+}}$, m_{Cl^-}). Then use the condition of electroneutrality to eliminate m_{Cl^-} and thereby to express a_{MgCl_2} in terms of $\gamma_\pm$ and $m_{Mg^{2+}}$ only.

2. The above expression is applicable to describe the thermodynamic state of an aqueous MgCl$_2$ solution (side β) in membrane equilibrium with a mixture of the Mg salt of DNA and MgCl$_2$ (side α). On side α, however, the mean ionic activity coefficient ($\gamma_{\pm,\alpha}$) and the ion concentrations ($m_{Mg^{2+},\alpha}$, $m_{Cl^-,\alpha}$) are affected by the presence of the DNA polyanion (present at the monomer [phosphate] concentration m_2). Use the condition of electroneutrality on side α to express $a_{MgCl_2,\alpha}$ in terms of $\gamma_{\pm,\alpha}$, $m_{Mg^{2+},\alpha}$, and m_2 only.

3. Start from the *general* condition for membrane equilibrium in terms of the *chemical potentials* of MgCl$_2$ on sides α and β and derive the *general* relationship between $m_{Mg^{2+},\alpha}$ and $m_{Mg^{2+},\beta}$ at membrane equilibrium in terms of m_2 and the activity coefficient ratio $\gamma_{32} \equiv \gamma_{\pm,\alpha}/\gamma_{\pm,\beta}$. *Neglect osmotic pressure effects.*

4. For the case of excess MgCl$_2$ ($m_{Mg^{2+},\alpha} \gg m_2$) obtain the *ideal* Donnan distribution of Mg^{2+} ($\gamma_{32} = 1$). Show that

$$\Gamma_{Mg^{2+}} \equiv \frac{m_{Mg^{2+},\alpha} - m_{Mg^{2+},\beta}}{m_2} = \frac{1}{3}$$

(*Hint:* Recall that $(1-x)^n \simeq 1 - nx$ for $0 < x \ll 1$.)

5. The *ideal* distribution coefficients of Mg^{2+} and Cl$^-$ at Donnan equilibrium are

$$\Gamma_{Mg^{2+}} = \frac{1}{3} \text{ and } \Gamma_{Cl^-} = -\frac{1}{3}$$

a. Interpret these ideal coefficients briefly. What basic thermodynamic principle dictates the ideal distribution of positive and negative electrolyte ions across the Donnan membrane?

b. The actual distribution coefficients in a MgCl$_2$–DNA Donnan equilibrium are $\Gamma_{Mg^{2+}} = 0.49$ and $\Gamma_{Cl^-} = -0.02$. Interpret these coefficients. What do they tell you about the actual equilibrium distribution of positive and negative electrolyte ions between side α and side β, and about the distribution of ions around each DNA molecule? How important is γ_{32} in determining the actual (as opposed to the ideal) ion distribution? Why?

6. In equilibrium dialysis experiments performed in excess NaCl to investigate the extent of interaction of Mg^{2+} with DNA, it is found that the Mg^{2+} distribution coefficient $\Gamma_{Mg^{2+}}$ (defined as above) decreases markedly as the NaCl concentration is increased at a fixed concentration of $MgCl_2$, such that $\Gamma_{Mg^{2+}} \to 0$ as $[NaCl] \to \infty$. Explain this observation at the level of the bulk solution *and* at the molecular level; i.e., what is the probable effect of adding NaCl on the distribution of Mg^{2+} around an individual DNA molecule. Why?

References

(1) Ingraham, J. L.; Maaloe, O.; Neidhardt, F. C. *Growth of the Bacterial Cell*; Sinauer Associates: Sunderland, Mass., 1983.
(2) Hvidt, A. *Ann. Rev. Biophys. Bioeng.* **1983**, *12*, 1–20.
(3) Record, M. T., Jr.; Lohman, T. M.; deHaseth, P. L. *J. Mol. Biol.* **1976**, *107*, 145–58.
(4) Tanford, C. *The Hydrophobic Effect*; Wiley: New York, 1976.
(5) Gill, S. J.; Nichols, N. F.; Wadso, I. *J. Chem. Thermodyn.* **1976**, *8*, 445–52.
(6) Richards, F. M. *Ann. Rev. Biophys. Bioeng.* **1977**, *6*, 151–76.
(7) Gill, S. J.; Dec., S. F.; Olofsson, G.; Wadso, I. *J. Phys. Chem.* **1985**, *89*, 3758–61.
(8) Edsall, J. T.; McKenzie, H. A. *Adv. Biophys.* **1978**, *10*, 137–207.
(9) Privalov, P. L. *Adv. Prot. Chem.* **1979**, *33*, 167–241.
(10) Privalov, P. L.; Griko, Y. V.; Venyaminov, S. Y. *J. Mol. Biol.* **1986**, *190*, 487–98.
(11) Schellman, J. A. *Ann. Rev. Biophys. Biophys. Chem.* **1987**, *16*, 115–38.
(12) Josephs, R.; Harrington, W. F. *Biochemistry* **1968**, *7*, 2834–47.
(13) Roe, J. H.; Burgess, R. R.; Record, M. T., Jr. *J. Mol. Biol.* **1985**, *184*, 441–53.
(14) Ray, A.; Reynolds, J. A.; Polet, H.; Steinhardt, J. *Biochemistry* **1966**, *5*, 2606–16.
(15) Hinz, H. J.; Weber, K.; Flossdorf, J.; Kula, M. R. *Eur. J. Biochem.* **1976**, *71*, 437–42.
(16) Suurkuusk, J.; Wadso, I. *Eur. J. Biochem.* **1972**, *28*, 438–41.
(17) Wishnia, A. *J. Phys. Chem.* **1963**, *67*, 2079–82.
(18) Mills, P.; Anderson, C. F.; Record, M. T., Jr. *J. Phys. Chem.* **1986**, *90*, 6541–48.
(19) Eisenberg, H. *Biological Macromolecules and Polyelectrolytes in Solution*; Clarendon Press: Oxford, 1976; p. 42.
(20) Strauss, V. P.; Helfgott, C.; Pink, H. *J. Phys. Chem.* **1967**, *71*, 2550–56.
(21) Anderson, C. F.; Record, M. T., Jr. *Biophys. Chem.* **1980**, *11*, 353–60.
(22) Anderson, C. F.; Record, M. T., Jr. In *Structure and Dynamics: Nucleic Acids and Proteins*; Clementi, E.; Sarma, R. H., Eds.; Adenine Press: New York, 1983; pp. 301–18.
(23) Manning, G. S. *J. Chem. Phys.* **1969**, *51*, 924–33.
(24) Stigter, D. *J. Phys. Chem.* **1978**, *82*, 1603–6.
(25) Manning, G. S. *Q. Rev. Biophys.* **1978**, *11*, 179–246.
(26) Anderson, C. F.; Record, M. T., Jr. *Ann. Rev. Phys. Chem.* **1982**, *33*, 191–222.
(27) Record, M. T., Jr. In *Unusual DNA Structures*; Wells, R.; Harvey, S., Eds.; Springer-Verlag: New York, 1987, pp. 237–51.
(28) Record, M. T., Jr.; Anderson, C. F.; Mills, P.; Mossing, M; Roe, J. H. *Adv. Biophys.* **1985**, *20*, 109–35.
(29) Record, M. T., Jr.; Anderson, C. F.; Lohman, T. M. *Quart. Rev. Biophys.* **1978**, *11*, 103–78.
(30) Kresheck, G. C.; Schneider, H.; Scheraga, H. A. *J. Phys. Chem.* **1965**, *69*, 3132–44.

Address correspondence to M. Thomas Record, Jr., at the Department of Chemistry, University of Wisconsin—Madison.

Chapter 10
COMPUTER METHODS IN THE CALCULATION OF PHASE AND CHEMICAL EQUILIBRIA

STANLEY M. WALAS
Department of Chemical & Petroleum Engineering
University of Kansas
Lawrence, Kan.

Methods for the prediction of physical, transport, and thermodynamic properties often employ formulas that are complex enough to warrant use of a computer, particularly when repeated evaluations are to be made. A definitive compilation of such methods is in progress under the sponsorship of the American Institute of Chemical Engineers (1) (1983 to present). They utilize more or less extensive data bases that are conveniently located in a computer and are often complex in execution, but usually they are direct methods in that no trial calculations are needed. Computer evaluations of many physicochemical problems, including the simpler relations of quantum and statistical mechanics, are described in the educational literature, for example, in the book by Benedek and Olti (2). The present chapter deals with phase and chemical equilibria of nonideal substances and mixtures whose evaluation requires the kinds of iterative calculations for which the computer is eminently suitable. A more nearly complete treatment is in Walas (3).

The Newton–Raphson method

Almost all realistic problems require numerical solution of complex equations. For this purpose, the Newton–Raphson method has been adopted in all of the examples of this chapter. When x_o is an approximate root of the single equation

$$f(x) = 0 \tag{1}$$

a correction in terms of the function and its derivative,

$$h = -f(x_o)/f'(x_o) \tag{2}$$

can be evaluated to make an improved value of the root

$$x = x_o + h = x_o - f(x_o)/f'(x_o) \tag{3}$$

The process is repeated to convergence with any desired tolerance.

When the analytical form of the derivative is difficult to find, a numerical evaluation with the computer is feasible, for example,

$$f'(x_o) = \frac{f(1.0001\ x_o) - f(x_o)}{0.0001\ x_o} \tag{4}$$

which makes the correction

$$h = -\frac{0.0001\ x_o f(x_o)}{f(1.0001\ x_o) - f(x_o)} \tag{5}$$

In Example 1 at the end of this chapter, the equation is $f(z) = z^3 - 0.9575\ z^2 + 0.18928\ z - 0.010004 = 0$. For a trial value $z_o = 0.7$, the correction is $h = 0.0001(0.7)(-0.0036832)/(-0.00366088 + 0.0036832) = 0.01155$. Accordingly, the first improved value is $z = 0.71155$, compared with the correct value 0.71110.

Convergence sometimes is facilitated by reducing the calculated correction by an arbitrary factor, thus converting equation 2 into the form

$$x = x_o + h/n \tag{6}$$

where n is a factor such as 5 or 10. Several of the Examples at the end of this chapter employ such a modification.

Simultaneous equations in several variables may also be solved by the Newton–Raphson method. The procedure is explained in the section on liquid–liquid equilibria and is illustrated by Example 5.

Equations of state

A key relation in applications of thermodynamics is that among pressure, temperature, and volume, called an equation of state (EOS), of which the simplest form is the ideal gas equation, $PV = RT$. Of the numerous EOS that have been proposed and are being actively used, the Peng–Robinson equation is adopted for illustration in this chapter. It has the merits of relative simplicity, good accuracy over a useful range of conditions, applicability to mixtures, and even modest accuracy in representing some kinds of liquid phase behavior.

Formulas relating to this equation are summarized in Table I. For pure substances, the properties that must be known are the critical pressure, the critical temperature, and the acentric factor. For highest accuracy with mixtures, empirical binary interaction factors also should be known for each pair of substances concerned. Unfortunately, interaction parameter data are not abundant in the open literature, although a few are cited in Table I. For other situations, at least an approximate representation of the PVT behavior of mixtures is obtained by taking the interaction parameters to be 0.

The Peng–Robinson equation is cubic in the volume or in the compressibility factor. Although an analytical solution for the roots of cubic equations is known, usually it is more convenient to find the real ones by the Newton–Raphson method. When all of the roots are real, the largest value of z is for the gas phase and the smallest for the liquid phase; the intermediate root is without physical significance. The equation in terms of compressibility, equation 7 of Table I, is most convenient for numerical evaluation because the range of z is small. A trial value $z = 1$ almost invariably converges to the largest root, and a trial value of about $z = 0.05$ converges to the smallest root. Example 1 is a computer solution of such a problem.

Enthalpy and entropy

When substances do not exhibit ideal gas behavior, changes in their enthalpies and entropies with changes of pressure may be evaluated with the aid of an EOS. The effects of changes in both T and P are represented by

---

Table I. The Peng–Robinson equation of state for pure substances and for mixtures

Standard form:

$$P = \frac{RT}{V - b} - \frac{a\alpha}{V^2 + 2bV - b^2} \tag{1}$$

Parameters:

$$a = 0.45724 R^2 T_c^2 / P_c \tag{2}$$

$$b = 0.07780 RT_c / P_c \tag{3}$$

$$\alpha = [1 + (0.37464 + 1.54226\omega - 0.26992\omega^2)(1 - T_r^{0.5})]^2 \tag{4}$$

$$A = a\alpha P / R^2 T^2 = 0.45724 \alpha P_r / T_r^2 \tag{5}$$

$$B = bP/RT = 0.07780 P_r / T_r \tag{6}$$

Polynomial form:

$$z^3 - (1 - B)z^2 + (A - 3B^2 - 2B)z - (AB - B^2 - B^3) = 0 \tag{7}$$

Mixtures:

$$a\alpha = \sum\sum y_i y_j (a\alpha)_{ij} \tag{8}$$

$$b = \sum y_i b_i \tag{9}$$

$$(a\alpha)_{ij} = (1 - k_{ij})\sqrt{(a\alpha)_i (a\alpha)_j} \tag{10}$$

$$A = \sum\sum y_i y_j A_{ij} \tag{11}$$

$$B = \sum y_i B_i \tag{12}$$

$$A_{ij} = (1 - k_{ij})(A_i A_j)^{0.5} \tag{13}$$

$$k_{ii} = 0 \tag{14}$$

	$k_{ij} =$
nitrogen + HC	0.12
CO_2 + HC	0.15
ethane + HC	0.01
propane + HC	0.01

$$H_2 - H_1 = \int_{T_1}^{T_2} C_p^* \, dT + \int_{P_1}^{P_2} (\delta H/\delta P)_T \, dP \qquad (7)$$

$$S_2 - S_1 = \int_{T_1}^{T_2} (C_p^*/T) \, dT - R \ln(P_2/P_1) + \int_{V_1}^{V_2} (\delta S/\delta V)_T \, dV \qquad (8)$$

where C_p^* is the heat capacity in the ideal gas state or at zero pressure.

In practical applications, it is convenient to standardize residual enthalpy $\Delta H'$ and residual entropy $\Delta S'$ at the system temperature as the difference between the value of the ideal property (at zero pressure) and the value at the system pressure. Thus,

$$\Delta H' = H^{id} - H = \int_0^P (\delta H/\delta P)_T \, dP \qquad (9)$$

$$= RT - PV + \int_\infty^V [P - T(\delta P/\delta T)_V] \, dV \qquad (10)$$

$$\Delta S' = S^{id} - S = \int_\infty^V [R/V - (\delta P/\delta T)_V] \, dV - R \ln(PV/RT) \qquad (11)$$

After the derivatives have been evaluated from the EOS, the integration can be completed. In terms of the residual properties, the changes in enthalpy and entropy between two states may be written

$$H_2 - H_1 = \int_{T_1}^{T_2} C_p^* \, dT + \Delta H_1' - \Delta H_2' \qquad (12)$$

$$S_2 - S_1 = \int_{T_1}^{T_2} (C_p^*/T) \, dT - R \ln(P_2/P_1) + \Delta S_1' - \Delta S_2' \qquad (13)$$

Formulas for the residual properties in terms of the Peng–Robinson equation are in Table II.

Table II. Residual enthalpy and entropy and fugacity coefficients from the Peng–Robinson equation of state

$$D_i = -T \frac{d(a\alpha)_i}{dT} = [m(a\alpha)\sqrt{T_r/\alpha}]_i \qquad (1)$$

$$D = \sum_i \sum_j y_i y_j m_j (1 - k_{ij}) \sqrt{a_i \alpha_i} \sqrt{a_j T_{rj}} \qquad (2)$$

$$\frac{\Delta H'}{RT} = 1 - z + \frac{A}{2.828\,B}\left(1 + \frac{D}{a\alpha}\right) \ln \frac{z + 2.414\,B}{z - 0.414\,B} \qquad (3)$$

$$\frac{\Delta S'}{R} = -\ln(z - B) + \frac{BD}{2.828\,A a\alpha} \ln \frac{z + 2.414\,B}{z - 0.414\,B} \qquad (4)$$

For pure substances, use D_i instead of D

$$\ln \phi = z - 1 - \ln(z - B) - \frac{A}{2\sqrt{2}B} \ln\left(\frac{z + 2.414\,B}{z - 0.414\,B}\right) \qquad (5)$$

$$\ln \hat{\phi}_i = \frac{B_i}{B}(z - 1) - \ln(z - B) + \frac{A}{4.828\,B}\left[\frac{B_i}{B} - \frac{2}{a\alpha} \sum_j y_j (a\alpha)_{ij}\right] \ln\left[\frac{z + 2.414\,B}{z - 0.414\,B}\right] \qquad (6)$$

where
  D_i = auxiliary parameter for pure substance i
  D = auxiliary parameter for a mixture
  $\Delta H'$ = residual enthalpy of a mixture
  $\Delta S'$ = residual entropy of a mixture
  ϕ = fugacity coefficient of a pure substance
  $\hat{\phi}_i$ = partial fugacity coefficient of component i of a mixture

Evaluations from several EOS are recorded in the literature in the forms of tables and graphs that are convenient for occasional use. Example 2 develops a computer program for evaluation of $\Delta H'$ of a ternary mixture with the P–R equation; the iterative solution for the compressibility factor z that is needed for this evaluation is included in the program.

A typical application of residual properties is calculation of expansion to a specified pressure in an adiabatic turbine with known efficiency

$$\eta = \Delta H/(\Delta H)_{\text{isentropic}} \tag{14}$$

The final isentropic temperature T_{2s} is first found by trial from

$$\Delta S = S_2 - S_1 = 0 = \int_{T_1}^{T_{2s}} (C_p^*/T)\, dT - \text{R} \ln (P_2/P_1) + \Delta S_1' - \Delta S_{2s}' \tag{15}$$

Note that the upper limit of integration and $\Delta S_{2s}'$ both depend on the unknown T_{2s}. After that has been found, the isentropic enthalpy change is found from

$$(\Delta H)_{\text{isentropic}} = \int_{T_1}^{T_{2s}} C_p^*\, dT + \Delta H_1' - \Delta H_{2s}' \tag{16}$$

The true enthalpy change is found by application of the known efficiency and is represented by

$$\Delta H = (\Delta H)_{\text{isentropic}}/\eta \tag{17}$$

$$= \int_{T_1}^{T_2} C_p^*\, dT + \Delta H_1' - \Delta H_2' \tag{18}$$

The final temperature T_2 is found by trial from equation 18; note that both the upper limit of integration and the value of $\Delta H_2'$ depend on T_2.

Phase equilibria

The basic condition for equilibrium to exist between phases is for the partial fugacities of individual components to be the same in all the phases. In vapor–liquid equilibria, for instance,

$$\hat{f}_i^{(V)} = \hat{f}_i^{(L)}, \text{ for every component } i \tag{19}$$

The fugacity is derivable from the EOS. For a pure substance the relation is

$$\ln (f/P) = \ln \phi = \frac{1}{\text{R}T} \int_0^P (V - \text{R}T/P)\, dP \tag{20}$$

where

$$\phi = f/P \tag{21}$$

is the fugacity coefficient. Similarly, the partial fugacity coefficient of a component i of a mixture is the ratio of the partial fugacity to the partial pressure,

$$\hat{\phi}_i = \hat{f}_i/y_i P \tag{22}$$

It is represented by

$$\ln \hat{\phi}_i = \frac{1}{\text{R}T} \int_0^P (\bar{V}_i - \text{R}T/P)\, dP \tag{23}$$

$$= \int_0^P [(\bar{z}_i - 1)/P]\, dP \tag{24}$$

The partial molal volume $\bar{V}_i$ and the partial compressibility $\bar{z}_i$ are defined by

$$\bar{V}_i = (\delta V/\delta n_i)_{T,P,n_j} \tag{25}$$

$$\bar{z}_i = (\delta z/\delta n_i)_{T,P,n_j} \tag{26}$$

The integrals of Equations 20 and 23 or 24 may be evaluated for known EOS. Table II records the results with the Peng–Robinson equation.

Vapor–liquid equilibria

Although the Peng–Robinson equation represents the behavior of liquid phases less accurately than the behavior of vapor phases, sometimes it may be used for evaluating fugacity coefficients of both phases and applied to the evaluation of vapor–liquid equilibria. The same form of equation is used for each phase, but the results are different because the compressibility z that appears in the fugacity equation is different for each phase. The largest root of $f(z) = 0$ applies to the vapor phase, and the smallest root to the liquid phase.

Upon substitution of equation 22 into equation 19 for each phase and rearranging, the relation between the vapor and liquid compositions becomes

$$y_i = (\hat{\phi}_i^{(L)}/\hat{\phi}_i^{(V)}) x_i \tag{27}$$

$$= K_i x_i \tag{28}$$

where

$$K_i = \hat{\phi}_i^{(L)}/\hat{\phi}_i^{(V)} \tag{29}$$

are the vaporization equilibrium ratios (VER).

Three main kinds of vapor–liquid equilibrium calculations are of interest: bubble points, dew points, and flashes at some fixed condition such as (T, P) or enthalpy or entropy. The condition for the bubble point of a mixture of known liquid composition x_i is represented by

$$f(T \text{ or } P) = \sum K_i x_i - 1 = 0 \tag{30}$$

and that for the dew point with known vapor composition y_i by

$$f(T, P) = \sum y_i/K_i - 1 = 0 \tag{31}$$

A flash process at fixed T and P and with a known overall composition z_i is represented by

$$f(\beta) = -1 + \sum x_i = -1 + \sum \frac{z_i}{1 + \beta(K_i - 1)} = 0 \tag{32}$$

where

$$\beta = V/F \tag{33}$$

is the fraction of the original mixture that becomes vapor at equilibrium. The individual terms of the summation represent the equilibrium liquid compositions x_i and the individual vapor mole fractions by $y_i = K_i x_i$. Equation 32 also applies to flashes at fixed enthalpy or entropy, but complete representation of those processes requires additional equations.

The derivative of the flash equation,

$$f'(\beta) = \sum \frac{(K_i - 1)z_i}{[1 + \beta(K_i - 1)]^2} \tag{34}$$

is required for implementation of the Newton–Raphson method.

Before embarking on a flash calculation, it is advisable to establish first whether two phases really exist by finding the dew and bubble point conditions. An algorithm for making a flash calculation is given in Figure 1, and a similar one is given in Figure 2; the latter is applied in Example 3.

Activity coefficients

Since the PVT behavior of liquids is not well represented by EOS such as the Peng–Robinson equation, another approach has been developed for the partial fugacities, namely,

$$\hat{f}_i = \gamma_i x_i f_i^0 \tag{35}$$

where f_i^0 is the fugacity of the pure component and γ_i is called the activity coefficient at the T and P of the system. As a consequence, the vaporization equilibrium ratio becomes

$$K_i = y_i/x_i = \gamma_i \phi_i^{\text{sat}} P_i^{\text{sat}} (PF)_i / \hat{\phi}_i^{(V)} P \tag{36}$$

where $(PF)_i$ is a relatively small correction for pressure (the Poynting factor) that is usually ignored, and ϕ_i^{sat} and P_i^{sat} are the fugacity and vapor pressure of the pure substance i at the system T and P.

At pressures below 6–8 atm or so, the ratio of fugacity coefficients in equation 36 often is near unity, in which case the VER becomes simply

$$K_i = \gamma_i P_i^{\text{sat}}/P \tag{37}$$

Figure 1. Algorithm for flash at fixed T and P with the same equation of state for each phase. Sometimes the initial assumption of ideal behavior, $K_i = P_i^{sat}/P$, will result in a phase split and thereby will give starting estimates of x_i and y_i, thus obviating the composition assumptions of Box 3. Some other estimates of compositions in Box 3 also may be necessary at times.

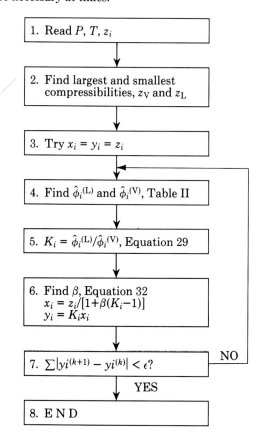

Figure 2. Algorithm for flash at fixed T and P with an equation of state for the vapor phase and activity coefficient correlations for the liquid state. As explained in connection with Figure 1, sometimes it may be necessary to estimate starting values of the phase compositions, when the assumptions of Box 2 do not lead to a phase split. Also it is advisable to check the dew and bubble point conditions before embarking on the flash calculation.

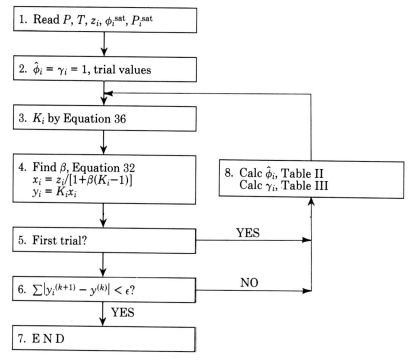

This approximate relation implies that the activity coefficient is a multiplicative correction to Raoult's law.

Activity coefficients depend on the nature and composition of the mixture, on the temperature, and to a limited extent on the pressure, although the last effect is commonly ignored. The dependence on composition can be correlated by a number of empirical equations. Those by Wilson and by Renon (the NRTL equation) are especially valuable because parameters evaluated from data of pairs of components often suffice to represent multicomponent behavior. Other equations in use also have this property, but this chapter will be restricted to the two mentioned.

Table III records these equations for two and any number of components. The Wilson equation is simpler in form, but it cannot represent liquid–liquid equilibria. The largest single compilation of parameters of activity coefficient correlations is a series of volumes on vapor–liquid and liquid–liquid systems published by DECHEMA (4).

An algorithm for a flash calculation employing an EOS for the vapor phase and activity coefficients for the liquid phase is given in Figure 2. Example 3 is an application of this procedure. Sometimes it is possible to start the iterative calculation with ideal vaporization equilibrium ratios,

$$K_i = P_i^{\text{sat}}/P \tag{38}$$

and thus to obtain initial estimates of the vapor and liquid compositions with which initial estimates of the fugacity and activity coefficients can be made. It often happens, however, that a phase separation is not obtained under ideal conditions when one actually exists. In such a case, start-up estimates must be made of the vapor and liquid compositions. Often the assumption that they are the same as the overall composition is satisfactory to begin with.

A useful type of calculation of vapor–liquid equilibria over a range of compositions and pressures is made in Example 4. At the low pressures involved, equation 37 is adequate. Again the Newton–Raphson method is employed for the iterative determination of the bubble point temperatures, and because of the complexity of the equations involved, the numerical form of the derivative, Equation 4, is adopted. Incidentally, this example illustrates the effect of pressure on the azeotropic composition of that mixture.

Liquid–liquid equilibria

When Equation 35 is written for each of the liquid phases in equilibrium, the condition for equality of partial fugacities becomes

$$\gamma_i x_i = \gamma_i^* x_i^* = \gamma_i^{**} x_i^{**} = \ldots, \text{ for all components } i \tag{39}$$

where the asterisks identify distinct liquid phases. The simplest case of only two components and two phases will be considered here.

Equation 39 may be rearranged into

$$p(x_1, x_1^*) = \ln(\gamma_i/\gamma_1^*) - \ln(x_1^*/x_1) = 0 \tag{40}$$

$$q(x_2, x_2^*) = \ln(\gamma_2/\gamma_2^*) - \ln(x_2^*/x_2) = 0 \tag{41}$$

with the requirements

$$x_1 + x_2 = x_1^* + x_2^* = 1 \tag{42}$$

Since the activity coefficients are functions of the mole fractions, by the Wilson or NRTL equations for instance, Equations 40 and 41 contain only the two phase compositions x_1 and x_1^* as unknowns, which may be found by simultaneous solution of the two equations.

The Newton–Raphson method can be applied to the solution of simultaneous equations in several variables. After estimates have been made of the compositions, corrections h and k can be developed to improve those estimates so that

$$x_1 = x_1 + h \tag{43}$$

$$x_1^* = x_1^* + k \tag{44}$$

The corrections are found by solution of the linear equations

$$hp_x + kp_{x^*} + p = 0 \qquad (45)$$

$$hq_x + kq_{x^*} + q = 0 \qquad (46)$$

with the results

$$h = -(pq_{x^*} - qp_{x^*})/(p_x q_{x^*} - p_{x^*} q_x) \qquad (47)$$

$$k = -(p_x q - q_x p)/(p_x q_{x^*} - p_{x^*} q_x) \qquad (48)$$

The four partial derivatives are

$$p_x = \delta p/\delta x_1 = \delta \ln \gamma_1/\delta x_1 + 1/x_1 \qquad (49)$$

Table III. Wilson and NRTL equations of activity coefficients

Binary mixtures

Name	Parameters	$\ln \gamma_1$ and γ_2
Wilson	λ_{12}	$-\ln(x_1 + \Lambda_{12}x_2) + x_2\left(\dfrac{\Lambda_{12}}{x_1 + \Lambda_{12}x_2} - \dfrac{\Lambda_{21}}{\Lambda_{21}x_1 + x_2}\right)$
	$\lambda_{21} - \lambda_{22}$	$-\ln(x_2 + \Lambda_{21}x_1) - x_1\left(\dfrac{\Lambda_{12}}{x_1 + \Lambda_{12}x_2} - \dfrac{\Lambda_{21}}{\Lambda_{21}x_1 + x_2}\right)$

$$\Lambda_{12} = \frac{V_2^L}{V_1^L} \exp\left(-\frac{\lambda_{12}}{RT}\right) \qquad \Lambda_{21} = \frac{V_1^L}{V_2^L} \exp\left(-\frac{\lambda_{21}}{RT}\right)$$

V_i^L molar volume of pure liquid component i.

NRTL	$g_{12} - g_{22}$ $g_{21} - g_{11}$	$x_2^2\left[\tau_{21}\left(\dfrac{G_{21}}{x_1 + x_2 G_{21}}\right)^2 + \dfrac{\tau_{12}G_{12}}{(x_2 + x_1 G_{12})^2}\right]$
	α_{12}	$x_1^2\left[\tau_{12}\left(\dfrac{G_{12}}{x_2 + x_1 G_{12}}\right)^2 + \dfrac{\tau_{21}G_{21}}{(x_1 + x_2 G_{21})^2}\right]$

$$\tau_{12} = \frac{g_{12} - g_{22}}{RT} \qquad \tau_{21} = \frac{g_{21} - g_{11}}{RT}$$
$$G_{12} = \exp(-\alpha_{12}\tau_{12}) \qquad G_{21} = \exp(-\alpha_{21}\tau_{21})$$

Multicomponent mixtures

Equation	Parameters	$\ln \gamma_i$
Wilson	$\Lambda_{ij} = \dfrac{V_j^L}{V_i^L}\exp\left(-\dfrac{\lambda_{ij}}{RT}\right)$ $\Lambda_{ii} = \Lambda_{jj} = 1$	$-\ln\left(\displaystyle\sum_{j=1}^m x_j \Lambda_{ij}\right) + 1 - \displaystyle\sum_{k=1}^m \dfrac{x_k \Lambda_{ki}}{\displaystyle\sum_{j=1}^m x_j \Lambda_{kj}}$
NRTL	$\tau_{ji} = \dfrac{(g_{ji} - g_{ii})}{RT}$ $G_{ji} = \exp(-\alpha_{ji}\tau_{ji})$ $\tau_{ii} = \tau_{jj} = 0$ $G_{ii} = G_{jj} = 1$	$\dfrac{\displaystyle\sum_{j=1}^m \tau_{ji} G_{ji} x_j}{\displaystyle\sum_{l=1}^m G_{li} x_l} + \displaystyle\sum_{j=1}^m \dfrac{x_j G_{ij}}{\displaystyle\sum_{l=1}^m G_{lj} x_l}\left(\tau_{ij} - \dfrac{\displaystyle\sum_{n=1}^m x_n \tau_{nj} G_{nj}}{\displaystyle\sum_{l=1}^m G_{lj} x_l}\right)$

$$p_x^* = \delta p/\delta x_1^* = \delta \ln \gamma_1^*/\delta x_1^* - 1/x_1^* \tag{50}$$

$$q_x = \delta q/\delta x_1 = \delta \ln \gamma_2/\delta x_1 - 1/(1-x_1) \tag{51}$$

$$q_{x^*} = \delta q/\delta x_1^* = \delta \ln \gamma_2^*/\delta x_1^* + 1/(1-x_1^*) \tag{52}$$

The derivatives of $\ln \gamma_1$ and $\ln \gamma_2$ of the NRTL equation are recorded in the computer program of Example 5.

The speed or even the possibility of convergence by the Newton–Raphson method is sensitive to the initial trial values of x_1 and x_1^*, so several sets may need to be tried. Reducing the magnitudes of the calculated correction factors by 5 is effective in forcing convergence in Example 5.

Melt equilibria

An equation for the solubility of a solid, originally proposed by Schröder (1893), is modified by inclusion of an activity coefficient into

$$\gamma x = \exp[(1 - T_m/T)\Delta S_m/R] \tag{53}$$

where

$$\Delta S_m = \Delta H_m/T_m \tag{54}$$

is the entropy of fusion or the ratio of the enthalpy of fusion and the temperature of fusion, and T is the system temperature. According to Walden's rule (1908), the entropy of fusion of organic compounds is approximately 13 cal/(gmol)(K).

The application made here is to the determination of the eutectic temperature and composition for which the condition is

$$\sum x_i = \sum \frac{1}{\gamma_i} \exp[(1 - T_{mi}/T)\Delta S_{mi}/R] = 1 \tag{55}$$

The activity coefficients are functions of the unknown eutectic temperature T and the eutectic compositions x_i.

A binary eutectic condition is evaluated in Example 6 with the aid of the Wilson equation. The procedure consists of these steps:

1. Initially assume that $\gamma_1 = \gamma_2 = 1$
2. Find the temperature by trial from

$$f(T) = (1/\gamma_1)\exp((1 - T_{m1}/T)\Delta S_{m1}/R) + (1/\gamma_2)\exp((1 - T_{m2}/T)\Delta S_{m2}/R) - 1 = 0$$

3. The exponential terms individually are x_1 and x_2.
4. With this information, first estimates of the activity coefficients are evaluated.
5. Substitution into the equation of Step 2 enables improved estimates of the eutectic condition, from which new values of the activity coefficients are found.
6. The process is continued to convergence.

The method is extended readily to any number of components.

Relaxation method for chemical equilibria

Equilibrium of several simultaneous chemical reactions is a condition of minimum Gibbs energy. The applicable equations usually are complex enough to require solution by iterative processes. Moreover, the minimum is a constrained one, because at each step of the calculation the amounts of individual elements must be the same as they are at the beginning. A method applicable to constrained minima is based on Lagrange multipliers. Unfortunately that procedure is too lengthy for inclusion in this chapter, but it is described completely in the literature, for example, by Smith and Missen (5), who provide algorithms and computer programs.

An alternate procedure that is feasible when the number of independent reactions is not more than five or six is the method of relaxation. Since each chemical reaction is represented by an equilibrium equation with an individual degree of advancement, the equilibrium equations could be solved simultaneously. Because the numerical solution of a system of nonlinear equations is a laborious process, the relaxation method replaces that process by a series of solutions of equations in single variables, one at a time, although those equations need to be solved several times before convergence is obtained.

The basis of the method is represented by Figure 3. Each reaction is assumed to proceed to equilibrium independently in its own vessel. The amounts of participants resulting from previous reactions serve as starting amounts for the reaction under consideration. A closed loop is formed. The equilibrium equations are written in terms of individual degrees of advancement that eventually go to 0. Example 7 illustrates this method with a two-reaction system.

Figure 3. Diagram of the relaxation or series-reactor method for equilibrium of multiple reactions. Each reactor operates under batch conditions and attains equilibrium of only its numbered reaction. The process is started by charging the starting mixture to reactor 1 where reaction 1 attains equilibrium. That product is transferred to reactor 2 where only reaction 2 comes to equilibrium. Transfer of that product to reactor 3 is made, and equilibrium of reaction 3 is forced. Then the material is returned to reactor 1, and the process is continued until the effluent composition from any reactor becomes constant.

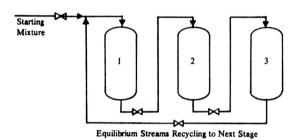

Example 1. Compressibility of a mixture with the Peng–Robinson equation. The mixture contains 40% CO_2 and 60% C_3H_6 and is at 303 K and 25.5 atm. The properties of the components and the mixture are tabulated along with calculations made with the formulas of Table I. R = 0.08205. $k_{12} = 0.15$.

	y	T_c	P_c	T_r	ω	a	b
CO_2	0.4	304.2	72.9	0.9961	0.225	3.9074	0.0266
C_3H_6	0.6	364.9	45.5	0.8304	0.148	9.0181	0.0513

	α	$a\alpha$	A	B
CO_2	1.0028	3.9183	0.1617	
C_3H_6	1.1088	9.9993	0.4125	
ij			0.2195	(cross value)
mixture			0.2797	0.0425

The values $A = 0.2797$ and $B = 0.0425$ are input to the program.

```
10 ! Example 1. Compressibility
     of a mixture with the P-R Eq
20 SHORT Z
30 READ A,B
40 DATA .2797,.0425
50 Z=1 ! Trial value
60 GOSUB 80
70 Z=.05 ! Trial value
80 ! Subroutine 1
90 GOSUB 190
100 F1=F
110 Z=1.0001*Z
120 GOSUB 190
130 F2=F
140 H=.0001*Z*F1/(F2-F1)
150 Z=Z/1.0001-H
160 IF ABS(H/Z)>=.0001 THEN 60
170 PRINT "Z=";Z
180 RETURN
190 ! Subroutine 2
200 F=Z^3-(1-B)*Z^2+(A-3*B^2-2*B
    )*Z-(A*B-B^2-B^3)
210 RETURN

The compressibilities are
   Z=0.7111, vapor phase
     0.15650, redundant
     0.089895, liquid phase
```

Note on the computer language employed: The Examples are written in Hewlett-Packard BASIC for the HP-85 computer.

Example 2. Residual properties with the Peng–Robinson equation.

The residual enthalpy of an equimolar mixture of ethane, propane, and n-butane will be found at 30 atm and 478 K (400 °F). Equation 3 of Table II is applicable. The required value of the compressibility is found with the Newton–Raphson method, but in terms of the analytical form of the derivative, line 460, instead of the numerical form employed in Example 1.

```
10  ! Example 2.  Residual enth
    alpy from the P-R eq, Eq 18
     of table 1.
15  SHORT Z,H
20  ! T1=Tc, P1=Pc, A1=alpha,W=o
    mega
30  OPTION BASE 1
40  DIM A(3),B(3),M(3),W(3),T1(3
    ),P1(3),Y(3),A1(3)
50  N=3
60  READ T,P,R
70  DATA 478,30,.08205
80  MAT INPUT Y
90  MAT READ T1,P1,W
100 DATA 305.4,369.8,425.2,48.2,
    41.9,37.5,.098,.132,.193
110 GOSUB 280
120 INPUT Z ! Trial value
130 GOSUB 440
140 H=F/F1
150 Z=Z-H
160 IF ABS(H/Z)<=.0001 THEN 180
170 GOTO 130
180 D=0
190 FOR I=1 TO N
200 FOR J=1 TO N
210 D=D+Y(I)*Y(J)*(A(I)*A(J))^.5
    *(1+.5*M(I)*(T/A1(I)/T1(I))^
    .5+.5*M(J)*(T/A1(J)/T1(J))^.
    5)
220 NEXT J
230 NEXT I
240 H=1.987*T*(1-Z+A/B/8^.5*(1+D
    *P/A/(R*T)^2)*LOG((Z+2.414*B
    )/(Z-.414*B)))

245 PRINT "Y1=Y2=Y3=1/3"
250 PRINT "Z=";Z
260 PRINT "ΔH'cal/gmol=";H
270 END
280 ! Subroutine for A and B
290 B=0
300 A=0
310 FOR I=1 TO N
320 M(I)=.3746+1.54226*W(I)-.269
    92*W(I)^2
330 A1(I)=(1+M(I)*(1-(T/T1(I))^.
    5))^2
340 A(I)=.45724*A1(I)*P/P1(I)*(T
    1(I)/T)^2
350 B(I)=.0778*P/P1(I)*T1(I)/T
360 B=B+Y(I)*B(I)
370 NEXT I
380 FOR I=1 TO N
390 FOR J=1 TO N
400 A=A+Y(I)*Y(J)*(A(I)*A(J))^.5
410 NEXT J
420 NEXT I
430 RETURN
440 ! Subroutine for Z
450 F=Z^3-(1-B)*Z^2+(A-3*B^2-2*B
    )*Z-(A*B-B^2-B^3)
460 F1=3*Z^2-2*(1-B)*Z+(A-3*B^2-
    2*B)
470 RETURN
```

```
Y1=Y2=Y3=1/3
Z= .87986
ΔH'cal/gmol= 290.4
```

Example 3. Flash at fixed T and P with the Peng–Robinson and Wilson equations.

The mixture is acetone/water with 20% of the former, at 210 °C and 30 bar. Vapor pressures are $P_1^0 = 29.71$ bar and $P_2^0 = 19.18$ bar. The Wilson parameters are $\Lambda_{12} = 0.05330$ and $\Lambda_{21} = 0.62313$. The binary interaction parameter is assumed to be 0. Other properties and derived data are tabulated.

	ω	T_c	P_c	a	b	α	$a\alpha$	A	B
acetone	0.309	508.1	49.0	16.1852	0.0662	1.0414	16.8548	0.3753	0.0584
water	0.344	647.3	220.5	5.8493	0.0187	1.2516	7.3211	0.1630	0.0165

The analytical derivative of the flash equation is employed in the Newton–Raphson procedure; the calculated increment is reduced by a factor of 20 (line 230). Usually it is advisable to check dew and bubble points before embarking on a flash calculation.

```
10  ! Example 3. Flash at fixed
     T and P with the P-R & Wilso
    n Eqs.
20  SHORT V,X1,Y1,F,F1,F2,G1,G2
30  ! Discontinue run when the d
    isplayed V becomes constant
40  READ T,P,Z1,Z2
50  DATA 483.2,30,.2,.8
60  READ P1,P2,L1,L2
70  DATA 29.71,19.18,.0533,.6231
    3
80  READ A1,A2,B1,B2,C1,C2,C3
90  DATA .3753,.163,.0584,.0165,
    16.8548,7.3211,11.1084
100 READ X1,X2,Y1,Y2
110 DATA .2,.8,.5,.5
120 GOSUB 270 ! activity coeffs
130 GOSUB 320 ! P-R params
140 Z=1
150 GOSUB 420 ! compressibility
160 Z=Z-H4
170 IF ABS(H4/Z)>=.0001 THEN 150
180 GOSUB 370 ! fugacity coeffs

190 K1=G1*F*P1/F1/P
200 K2=G2*F*P2/F2/P
210 V=.4
220 GOSUB 480 ! fraction vapor V
230 V=V-J2/20
240 IF ABS(J2/V)>=.005 THEN 220
250 DISP V
260 GOTO 120
270 ! Subroutine for activity co
    effs
280 D=L1/(X1+X2*L1)-L2/(X1*L2+X2
    )
290 G1=EXP(-LOG(X1+X2*L1)+X2*D)
300 G2=EXP(-LOG(X1*L2+X2)-X1*D)
310 RETURN
320 ! SR for P-R parameters
330 A=Y1^2*A1+Y2^2*A2+2*Y1*Y2*(A
    1*A2)^.5
340 B=Y1*B1+Y2*B2
350 C=Y1^2*C1+Y2^2*C2+2*Y1*Y2*C3
360 RETURN
370 ! SR for fugacity coeffs
```

```
380 F=EXP(Z-1-LOG(Z-B)-A/8^.5/B*
    LOG((Z+2.414*B)/(Z-.414*B)))
390 F1=EXP(B1/B*(Z-1)-LOG(Z-B)+A
    /4.828/B*(B1/B-2/C*(Y1*C1+Y2
    *C3))*LOG((Z+2.414*B)/(Z-.41
    4*B)))
400 F2=EXP(B2/B*(Z-1)-LOG(Z-B)+A
    /4.828/B*(B2/B-2/C*(Y1*C3+Y2
    *C2))*LOG((Z+2.414*B)/(Z-.41
    4*B)))
410 RETURN
420 ! SR for compressibility Z
430 M=Z^3-(1-B)*Z^2+(A-3*B^2-2*B
    )*Z-(A-B-B^2-B^3)
440 M1=3*Z^2-2*(1-B)*Z+A-3*B^2-2
    *B
450 H4=M/M1
460 RETURN
470 END
480 ! SR for vapor fraction V
490 X1=Z1/(1+V*(K1-1))
500 X2=Z2/(1+V*(K2-1))
510 J=X1+X2-1
520 Y1=K1*X1
530 Y2=K2*X2
540 J1=X1*(K1-1)/(1+V*(K1-1))+X2
    *(K2-1)/(1+V*(K2-1))
550 J2=-(J/J1)
560 RETURN
570 END
```

```
Fraction vapor = .38248
X1= .066993, mol fraction liquid
Y1= .41421, mol fraction vapor
G1=7.1011, activity coeff of 1
G2=1.0392, activity coeff of 2
F= .81962, fugacity coeff of mix
F1=.93104, fugacity coeff of 1
F2=.88254, fugacity coeff of 2
Z=1.001856, compressibility of
           the mixture.
```

Example 4. Vapor–liquid equilibria of methyl acetate/methanol at several pressures, on the basis of the Wilson and Antoine equations, and neglecting vapor phase nonidealities.

```
10 ! Example 4. The x-y diagram
    at several pressures with W
   ilson Eq.
20 SCALE 0,1,0,1
30 XAXIS 0,.1
40 XAXIS 1,.1
50 YAXIS 0,.1
60 YAXIS 1,.1
70 P=1 ! Change to desired valu
   e
80 FOR X=0 TO 1 STEP .05
90 X2=1-X
100 T=350 ! Trial value
110 GOSUB 230
120 F1=F
130 T=1.0001*T
140 GOSUB 230
150 F2=F
160 H=.0001*T*F1/(F2-F1)
170 T=T/1.0001-H
180 IF ABS(H/T)>=.0001 THEN 110
190 Y=G1*P1*X/P
200 DRAW X,Y
210 NEXT X
220 END
230 ! Subroutine for temperature
240 A1=.5108*EXP(54.9958/T)
250 A2=1.9578*EXP(-(467.79/T))
260 P1=EXP(16.5835-2838.7/(T-45.
    16))/760
270 P2=EXP(18.1412-3391.96/(T-43
    .16))/760
280 G=A1/(X+A1*X2)-A2/(A2*X+X2)
290 G1=EXP(-LOG(X+A1*X2)+X2*G)
300 G2=EXP(-LOG(X2+A2*X)-X*G)
310 F=-P+G1*P1*X+G2*P2*X2
320 RETURN
```

Example 5. Liquid–liquid equilibrium compositions of nitromethane/heptane at 70 °C, with NRTL parameters from Reference 6.

Note: Trial values for (x_1, x_1^*) of (0.1, 0.8), (0.2, 0.8), (0.3, 0.9) and (0.4, 0.8) converge to (0.1378, 0.9539) with the full corrections h and k; the trial (0.3, 0.8) requires that h and k be divided by 5 at each iteration.

```
10 ! Example 5. Liquid-liquid e
   quilib of nitromethane/hepta
   ne @ 70C with the NRTL Eq.
20 READ T1,T2,G1,G2
30 DATA 2.5766,.9908,.5973,.820
   2
40 INPUT X1,Y1
50 X2=1-X1
60 Y2=1-Y1
70 GOSUB 220
80 D=P1*Q2-P2*Q1
90 H=(P*Q2-Q*P2)/D
100 K=(P1*Q-Q1*P)/D
110 DISP H;K
120 DISP P;Q
130 DISP
140 X1=X1-H ! divide H & K by 5
    to improve convergence if ne
    eded
150 Y1=Y1-K
160 X2=1-X1
170 Y2=1-Y1
180 IF ABS(H)+ABS(K)>=.0002 THEN
     70
190 PRINT "X1=";X1
200 PRINT "X1*=";Y1
210 END
220 ! Subroutine for corrections
     H & K
230 GOSUB 320
240 P1=U+1/X1
250 Q1=V-1/X2
260 GOSUB 400
270 P2=U-1/Y1
280 Q2=V+1/Y2
290 P=U1-V1-LOG(Y1/X1)
300 Q=U2-V2-LOG(Y2/X2)
310 GOTO 80
320 ! Subroutine for derivatives
330 U=-(2*X2*(T2*(G2/(X1+X2*G2))
    ^2+T1*G1/(X2+X1*G1)^2))
340 U=U+X2^2*(-(2*T2*G2^2*(1-G2)
    /(X1+X2*G2)^3)-2*T1*G1*(-1+G
    1)/(X2+X1*G1)^3)
```

```
350 V=2*X1*(T1*(G1/(X2+X1*G1))^2
    +T2*G2/(X1+X2*G2)^2)
360 V=V+X1^2*(2*T1*G1^2*(-1+G1)/
    (X2+X1*G1)^3-2*T2*G2*(1-G2)/
    (X1+X2*G2)^3)
370 U1=X2^2*(T2*(G2/(X1+X2*G2))^
    2+T1*G1/(X2+X1*G1)^2)
380 V1=X1^2*(T1*(G1/(X2+X1*G1))^
    2+T2*G2/(X1+X2*G2)^2)
390 RETURN
400 ! Subroutine for *phase
410 U=-(2*Y2*(T2*(G2/(Y1+Y2*G2))
    ^2+T1*G1/(Y2+Y1*G1)^2))
420 U=U+Y2^2*(-(2*T2*G2^2*(1-G2)
    /(Y1+Y2*G2)^3)-2*T1*G1*(-1+G
    1)/(Y2+Y1*G1)^3)
430 V=2*Y1*(T1*(G1/(Y2+Y1*G1))^2
    +T2*G2/(Y1+Y2*G2)^2)
440 V=V+Y1^2*(2*T1*G1^2*(-1+G1)/
    (Y2+Y1*G1)^3-2*T2*G2*(1-G2)/
    (Y1+Y2*G2)^3)
450 U2=Y2^2*(T2*(G2/(Y1+Y2*G2))^
    2+T1*G1/(Y2+Y1*G1)^2)
460 V2=Y1^2*(T1*(G1/(Y2+Y1*G1))^
    2+T2*G2/(Y1+Y2*G2)^2)
470 RETURN
```

Example 6. The eutectic conditions of a binary mixture with the aid of the Wilson equation.

The entropies of fusion are each 13 cal/gmol, the temperatures of fusion are 350 K and 425 K, and the Wilson parameters are $\Lambda_{12} = 1.25 \exp(-1166.2/T)$ and $\Lambda_{21} = 0.8 \exp(224.7/T)$. The results converge in three iterations.

```
10  ! Example 6. Eutectic temp a
    nd composition with the Schr
    oder & Wilson Eqs.
20  SHORT T,X1,G1,G2
30  READ G1,G2 ! Trial values
40  DATA 1,1
50  T=325 ! Trial value
60  GOSUB 280
70  F1=F
80  T=1.0001*T
90  GOSUB 280
100 F2=F
110 H=.0001*T*F1/(F2-F1)
120 T=T/1.0001-H
130 IF ABS(H/T)>=.0001 THEN 60
140 GOSUB 330
150 GOSUB 280
160 F1=F
170 T=1.0001*T
180 GOSUB 280
190 F2=F
200 H=.0001*T*F1/(F2-F1)
210 T=T/1.0001-H
220 IF ABS(H/T)>=.0001 THEN 150
230 PRINT "Eutectic temp=";T
240 PRINT "X1=";X1
250 PRINT "G1=";G1
260 PRINT "G2=";G2
270 END
280 ! Subroutine for T
290 X1=1/G1*EXP(6.5425*(1-350/T)
    )
300 X2=1/G2*EXP(6.5425*(1-425/T)
    )
310 F=-1+X1+X2
320 RETURN
330 ! Subroutine for activity co
    efficients G1, G2
340 L1=1.25*EXP(-(1166.2/T))
350 L2=.8*EXP(224.7/T)
360 B=L1/(X1+X2*L1)-L2/(X1*L2+X2
    )
370 G1=EXP(-LOG(X1+X2*L1)+X2*B)
380 G2=EXP(-LOG(X2+X1*L2)-X1*B)
390 RETURN
```

```
The solution is

Eutectic temperature = 342.92
X1= .8686, mol fraction of 1 in
     the eutectic
G1=1.0066, activity coeff of 1
G2=1.5822, activity coeff of 2
```

Example 7. Relaxation method for finding the equilibrium composition of the pair of reactions (1) $CH_4 + H_2O \rightleftarrows CO + 3H_2$ and (2) $CO + H_2O \rightleftarrows CO_2 + H_2$.

At 600 °C the equilibrium constants are $K_1 = 0.574$ and $K_2 = 2.21$. The participants are numbered $CH_4 = 1$, $CO_2 = 2$, $CO = 3$, $H_2O = 4$, $H_2 = 5$. The initial mixture contains 1 mol of CH_4 and 5 mol of H_2O. The equilibrium equations, on the assumption that each reaction proceeds independently but consecutively, are

$$f(e_1) = \frac{(n_3+e_1)(n_5+3e_1)^3 P^2}{(n_1-e_1)(n_4-e_1)(n+2e_1)^2} - 0.574 = 0$$

$$f(e_2) = \frac{(n_2+e_2)(n_5+e_2)}{(n_3-e_2)(n_4-e_2)} - 2.21 = 0$$

In each case the n_i are the moles present at the beginning of each step of the relaxation series of calculations. Eventually, the degrees of advancement e_1 and e_2 go to 0. Solutions are shown for $P = 1.1$ atm and 1.5 atm.

```
10  ! Relaxation method for find
    ing equilibria of multiple r
    eactions
20  SHORT N1,N2,N3,N4,N5,N
30  READ K1,K2
40  DATA .574,2.21
50  READ N1,N2,N3,N4,N5,P
60  DATA 1,0,0,5,0,1.1
70  READ E1,E2 ! Trial values
80  DATA .8,.5
90  GOSUB 450
100 F1=F
110 E1=1.0001*E1
120 GOSUB 450
130 F2=F
140 H1=.0001*E1*F1/(F2-F1)
150 E1=E1/1.0001-H1/5
160 IF ABS(H1/E1)>=.0001 THEN 90
170 N1=N1-E1
180 N3=N3+E1
190 N4=N4-E1
200 N5=N5+3*E1
210 N=N1+N2+N3+N4+N5
220 DISP N;E1
230 GOSUB 480
240 G1=G
250 E2=1.0001*E2
260 GOSUB 480
```

```
270 G2=G
280 H2=.0001*E2*G1/(G2-G1)
290 E2=E2/1.0001-H2/5
300 IF ABS(H2/E2)>=.0001 THEN 23
    0
310 N2=N2+E2
320 N3=N3-E2
330 N4=N4-E2
340 N5=N5+E2
350 N=N1+N2+N3+N4+N5
360 IF ABS(E1)+ABS(E2)>=.0002 TH
    EN 90
370 PRINT "N1=";N1
380 PRINT "N2=";N2
390 PRINT "N3=";N3
400 PRINT "N4=";N4
410 PRINT "N5=";N5
420 PRINT
430 PRINT "N=";N
440 END
450 ! Subroutine for reaction 1
460 F=-K1+(N3+E1)*(N5+3*E1)^3*P^
    2/(N1-E1)/(N4-E1)/(N1+N2+N3+
    N4+N5+2*E1)^2
470 RETURN
480 ! Subroutine for reaction 2
490 G=-K2+(N2+E2)*(N5+E2)/(N3-E2
    )/(N4-E2)
500 RETURN
510 END
```

```
P=1.1 atm            P = 1.5 atm

N1= .10009           N1= .14614
N2= .62772           N2= .60762
N3= .27219           N3= .24624
N4= 3.4724           N4= 3.5386
N5= 3.3275           N5= 3.1691

N= 7.7999            N= 7.7077
```

References

(1) *AIChE Manual for Predicting Chemical Process Design Data*; American Institute of Chemical Engineers, 1983 to present.
(2) Benedek, P.; Olti, F. *Computer-Aided Thermodynamics of Gases and Liquids*; Wiley: New York, 1985.
(3) Walas, S.M. *Phase Equilibria in Chemical Engineering*; Butterworths: Stoneham, Mass., 1985.
(4) Gmehling, J. et al. *Vapor–Liquid Equilibrium Data Collection*, DECHEMA, 1979 to present.
(5) Smith, W.R.; Missen, R.W. *Chemical Reaction Equilibrium Analysis Theory and Algorithms*, Wiley: New York, 1982.
(6) Sorenson, J.M.; Arlt, W. *Liquid–Liquid Equilibrium Data Collection*, DECHEMA, 1979–80.